Democracy and Environmental Movements in Eastern Europe

Democracy and Environmental Movements in Eastern Europe

A Comparative Study of Hungary and Russia

Katy Pickvance

Routledge
Taylor & Francis Group

LONDON AND NEW YORK

First published 1998 by Westview Press

Published 2018 by Routledge
52 Vanderbilt Avenue, New York, NY 10017
2 Park Square, Milton Park, Abingdon, Oxon OX14 4RN

Routledge is an imprint of the Taylor & Francis Group, an informa business

A CIP catalog record for this book is available from the Library of Congress.

ISBN 13: 978-0-367-01052-2 (hbk)
ISBN 13: 978-0-367-16039-5 (pbk)

Contents

Preface

The unexpected political changes in eastern Europe captured many people's imagination at the time. As I was born there, but lived in Great Britain, it was inevitable for me to turn my attention to my "home land". This book is the result of this academic u-turn.

In order to take on such a heroic task as to complete a book you need a lot of encouragement and mental support. I would like to thank those people who helped me in an indirect way to write this book.

I would like to dedicate this book to all these people.

First of all my mother, Gabi, whose pride in my achievements makes it worth all the effort and hard work. Secondly, my two kids, Balázs and Bence, whose mental support and love mean so much to me. Then, Chris, my husband, friend and intellectual partner, whose appreciation was very important to me during this work. Next, my cousin, Alex, who inspired me at the last stage of writing this book and made it easier for me to visualize 'the' intelligent reader for whom I was hoping to write this book. I would also like to thank very much András Hegedüs for being so supportive for many years both as a friend and as an intellectual influence.

Among my colleagues I would like to thank firstly, Chris Rootes, who helped me from the initial ideas to the final stage of this book. Secondly, Nick Manning, whose encouragement motivated me over many years, and our collaborators in the project, Sveta Klimova, Katalin Haskó, Lev Perepjolkin and Viktória Szirmai.

I would also like to thank the Economic and Social Research Council for financing the research on which this book is based.

Finally, but very importantly, I would like to thank Victor, our family dog, whose presence around me always cheered me up when I needed it and who provided wonderful excuses for a short break when demanding a little walk.

Katy Pickvance
Canterbury, UK, May 1998

1

Introduction

The end of the Communist regimes in eastern Europe has been described as a "revolution". Political systems which almost all writers considered to be stable and expected to continue for decades rapidly collapsed one after the other. Democracy was then expected to replace state socialist regimes which had allowed limited dissent. Likewise journalistic reporting described the probability of rapid transition from a state socialist to a capitalist economy as soon as the appropriate dose of western economic medicine had been taken. In other words the nature of what came after state socialism was "read off" from a mixture of ideological slogans and desires.

This book starts from a different place. As a Hungarian, who lived in Hungary until 1986, a student of Russian and a frequent visitor to the Soviet Union from 1970 onwards, I was aware of the gap between western images of state socialism and the reality of the Hungarian and Soviet systems. The purpose of this book is to go beyond the slogans and to study one aspect of reality of post-socialism in Hungary and Russia.

Hungary and Russia are particularly interesting to study because they had a very different historical development prior to the socialist period as well as during it. The recent regime change also took different forms in each case and even occurred in different years. In Hungary the gradual reform process resulted in a peaceful change in 1989 while in Russia there was a stormy regime change in 1991. Thus these two post-socialist societies provide a challenging comparison.

The choice of environmental movements reflects several factors. Environmentalism is one of the most important global currents of thinking today, and the extent of its presence in eastern Europe is of great interest. By definition environmental issues are, in part at least, intrinsically global in character. Hence eastern European reactions are of particular importance to people everywhere as well as being of academic interest. Secondly, it is well established that in eastern Europe environmental damage under state socialism was extremely serious. Hence the grounds for environmental activism were abundantly present.

The book is concerned with environmental movements in Hungary and Russia. However, since its aim is to offer a proper understanding of the differing patterns of environmental movement activity in the two countries, it gives considerable attention to political opposition under state socialism, and the historical experience (or absence) of democracy in the pre-socialist period. Only by understanding the long-term trends in the two societies can one understand present patterns.

The book has four aims. First it aims at exploring how far democracy developed in the two post-socialist societies. It does not, however, aim at staying on the surface by only looking at the present situation. The aim is to go deeper by analyzing past and present, to see whether there is any continuity and to what extent the past influences current democratic developments. The second aim is to present the results of an in-depth study of environmental movements in two eastern European societies and their relations with local and national authorities, as an example of grassroots activities in a democratic society. The third aim is to make a systematic comparison of the Russian and Hungarian situation. The final aim is to examine existing theories of opposition in Soviet-type societies, civil society theory and social movement theories to examine their relevance in the eastern European context.

The book thus has two major parts. Firstly, the discussion of democratic development (or the lack of it) prior to and during the socialist era and opposition in Soviet-type societies. The second part presents an examination of the empirical data, which is based on in-depth interviews conducted with social movement activists and leaders and local and national authority members, and the theoretical analysis.

Following the introduction the first part consists of three chapters (chapters 2, 3 and 4). Chapter 2 investigates the historical background of Hungary and chapter 3 that of Russia in order to find out whether either of these societies had any democratic experience in the pre-socialist period. Chapter 4 reviews the state of pluralism and opposition in Soviet-type regimes at a theoretical level and in practice, and it demonstrates that there was scope for opposition under socialism. The nature of this opposition was, however, very different in the two cases due to their different social and political character.

In the second part I analyse environmental movement activities and national and local authorities in the two post-socialist societies, based on our empirical research. Chapter 5 describes a number of environmental movements in Hungary including both those which came into existence in the mid 1980s and those which were founded in the early 1990s. The examples studied also provide a selection of local and national environmental movements, Budapest-centred and operating outside the capital, and the chapter examines these movements in order to explore their aims and goals, participants, leaders, leadership styles and internal conflicts, the role of the media and the degree and nature of success they achieved.

Next I turn to Russia. Chapter 6 describes the environmental movements in Russia which came into existence in the late 1980s and early 1990s. The examples studied provide a selection of national (federal) and local cases, of which some are Moscow based and others operate outside the capital. As in chapter 5 for Hungarian movements, I systematically examine the individual Russian cases in order to investigate their characteristics and actions.

The next two chapters focus on the relation between local and national authorities and environmental movements in Hungary (chapter 7) and in Russia (chapter 8). They examine the recent development of democratic institutions in Hungary and Russia both at local and national level. The aim of these chapters is to examine the extent of democratic development in order to establish why movements with many similar features in Hungary and Russia achieve such different degrees of success. It is argued that the main reasons for the success of environmental movements in Hungary, and the lack of it in Russia, lie in factors outside the movements themselves, namely the social and political context in which they are embedded. The movements' relation with the national and local authorities is the major focus of these two chapters.

The next chapter presents a comparison of Hungarian and Russian environmental movements which draws together the different aspects of the analysis explored in Chapters 5 to 8. Chapter 10 then examines the relevance of existing North American and West European theories in the eastern European context. This includes returning to the civil society theory and the "western" social movement theories, such as *collective behavior*, *resource mobilization*, *environmental consciousness*, *new social movement theories*, and the *political opportunity structure theory*, in order to apply them to the concrete cases of Russia and Hungary and examine to what extent they offer arguments applicable in the eastern European context. It is shown that most of the theories examined have some relevance to the cases considered although some were found more useful than others.

In the final chapter I draw an overall conclusion to the book, discussing its achievements in relation to its aims. I also explore the implications of the study for the development of social movement theories, and discuss the likely trends in social movement development and in the development of democratic institutions in the two societies studied specifically, and in eastern Europe generally.

2

"A Little Bit to the East from the West and a Little Bit to the West from the East": Democratic Development in Hungary

Introduction

One argument of this book is that the present state of democracy in Hungary and Russia is closely linked to the way democracy developed in these two societies historically. In other words, the contemporary development of democratic institutions (or the lack of it) is due to the continuity with the historical past. The lack of fully developed democratic institutions and practices predates the socialist period of these two countries (as in others). The key to their present differences therefore lies in many respects in the history of these two societies. In this chapter I am going to evaluate the Hungarian development throughout its history in order to find out whether there was any democratic presence there prior and during the socialist era. The following chapter will do the same in Russia.

It has been argued in the literature that, in relation to the development of democracy, there are three—rather than the usually quoted two—historical regions within Europe: western Europe, eastern Europe, and east-central Europe (Szűcs, 1988; Bibó, 1979). Hungary's history is most certainly better understood by this concept. It has always been influenced by the fact that it is located a little bit to the East from the West and a little bit to the West from the East. Partly because of this location and partly because of the path it has followed, Hungary became a semi-developed society over the centuries: more developed than its neighbours to the East (such as Russia) but less developed than its western European counterparts.

In contrast to Russia, Hungary developed more along western European lines. It took up Latin Christianity instead of Byzantine (Makkai, 1990) leading to a similar development to its contemporaries at the time in western Europe. The country's central location in Europe, however, resulted in a series of occupations over history both from the West and from the East.

Occupied Hungary: "Out of a Frying Pan into the Fire"

Shortly after "western type" Roman Christianity and feudalism had been established three major occupations occurred.[1] The last foreign rulers were the Habsburgs. In between two occupations the main trend of Hungarian development was towards a full adoption of western type institutions including a distinctly feudal class system (Engel, 1990), as was the case in many other European societies at the time. Other important factors were the Reformation and Counter-Reformation.[2] These occupations made the country weak and vulnerable. The Habsburg occupation looked more attractive from a religious point of view, as they were Christians as opposed to the Muslim Turks, but was a mixed blessing economically and politically.

Under the two hundred years of Habsburg influence Hungary both benefited and lost out. While it was not independent, some of the enlightened Habsburg policies improved the previous even more uneven distribution of wealth, political power and privileges. It led to an expansion of the education system which in turn resulted in changes in the political structure curtailing some of the traditionally strong privileges of the Hungarian nobility. Economically the Habsburg Empire was also a mixed blessing. It was beneficial in that it created a huge market for Hungarian agricultural products and helped to develop an industry and some infrastructure. This was, however, mainly agricultural-related industry (mills, for example) and manufacturing industry was allocated to other parts of the Empire. Another major problem was nationalism.

In the late eighteenth and early nineteenth century, under the influence of the European wave of Romanticism, all nations in Europe were looking for their "historical roots" to strengthen their national identity. The awakening Hungarian nationalism developed against the strong Habsburg influence but was blind to the parallel claims of Croats, Romanians or Slovaks. The leading members of society advocating narrowly Magyar-minded views were the lower nobility who engaged in frequent forceful Magyarization campaigns in the ethnic territories. Instead of introducing a pluralistic patriotism, it led to a narrow minded nationalistic pride (Bárány, 1990) despite the fact that less than 40 per cent of Hungarians were actually ethnic Hungarians[3], that is, only a minority of the population. In political terms, in the first half of the nineteenth century, some enlightened nobles, like Széchényi, were trying to introduce reforms such as the introduction of a modern party organization and freedom of speech in public meetings, following ideas borrowed from the British system (Bárány, 1990).

Another part of the opposition in the 1840s was a group of more radical lesser noblemen, led by Kossuth, open to western ideas and eager to challenge the absolutist Habsburg system. They supported the idea of a strong western type of parliamentary system, a government accountable to parliament, civil rights for all, including non-nobles, citizen representation in local, municipal and national legislative bodies, equality before the law, the abolition of serfdom, freedom of association, openness in public life, a free press, the lifting of censorship, general taxation, access to credit and the right to purchase land for all. But they did not mention the ethnic rights of non-Magyars (Barta, 1975; Bárány, 1990).

A third group presenting yet more radical opposition to the regime at the time, was a wing of young radicals of lesser noble intellectuals, admirers of the French Revolution, romantic utopians during the 1848-49 revolution which swept from Paris to Vienna, Prague, Milan and Budapest (Deme, 1976; Deák, 1990). Thus Hungary was by the middle of the nineteenth century very much in line with contemporary western European, mainly German, ideology but was still under Habsburg occupation (as a dual monarchy).

The Hungarian bureaucracy was surrounded with an aura of power and authority and was full of members of the declining nobility, the so-called gentry. "The best recommendation for an administrative position was a shining name" like Lipiczay, Eszterhazy or Dessewffy (Lippay, 1919, p.98). Non-Hungarian nationalities were accepted only if they were willing to assimilate, and an intensive "Magyarization" process was pursued (János, 1982; Seton-Watson, 1945). Also, a de facto regressive tax system had been maintained which forced smallholders to pay a 30 per cent tax rate while large estate owners had to pay only 9 per cent (Jeszenszky, 1990).

Several measures were introduced in the late nineteenth century, however, which pointed to more progressive political development: laws were passed to establish the basis for a modern bourgeois legal system with a judiciary, separate from the administration, and a system of independent judges (Hanák, 1984). Because the economy suffered from shortages of capital and of specialists (such as engineers, skilled workers, and entrepreneurs), these were welcomed from abroad, resulting in about 25 per cent of foreigners in the industrial labour force in Budapest and a mushrooming of industrial enterprises at the time. Enterprise and intensive market-oriented agriculture were strongly encouraged (Frank, 1990). Budapest became Europe's sixth largest city by the turn of century, with 1 million inhabitants excluding the suburbs (Höensch, 1988).

There were, however, serious consequences of this, somewhat delayed, development. While the proportion of the population living in urban areas had increased (Ranki, 1989; Cohen, 1989) the process of urbanization was almost a century behind western Europe (Hanák, 1984). The majority of the population remained tied to agriculture (Rothschild, 1988) and only a small proportion of the working class, which constituted only 13 per cent of the whole population, was skilled (Held, 1971). The new rising class was the bourgeoisie, the richest of whom were merchants, industrialists, bankers and a growing number of owners of small industrial enterprises, larger shops and apartment houses. At the lower end of the scale were the lesser artisans, owners of small shops and clerical staff.

The elite was still a small, closely knit group, owning the bulk of the land in the form of huge latifundia, thus maintaining their economic power, social prestige and feudalistic values (Deme, 1984). The "gentry", the class of empoverished small landowners and nobles, were a formidable conservative political and social force. They retained their privileged position and political influence through their extensive connections and by occupying key roles in the state bureaucracy. They maintained feudalistic values and looked down on the newly emerging bourgeoisie (Erdei, 1978) and the intelligentsia was tiny (4 per cent) (Karády, 1989; Frank, 1990).

Independent Hungary

By the end of World War I in 1918 the Habsburg Empire had become sufficiently weak to allow its central-eastern European parts to become independent societies. In the period between the two world wars power was held by extreme right wing politicians but they did not ban all democratic activity. Various conservative factions and pro-Habsburg royalists joined together to form the Christian National Unity Party. Other legal parties were the Independent Smallholders, uniting peasants; the National Democratic Party, representing the leftist section of the middle class; and the re-established Hungarian Social Democratic Party. Only the Communist party was outlawed from 1921 (Batkay, 1982; Nagy, 1975).

The period between the two world wars was not very successful in establishing and securing democratic civil rights. The regime remained right wing, anti-liberal, neo-conservative and nationalistic. Although it was not fascist prior to 1944, as some have claimed, the concept of western democracy did not exist either. The authority of Parliament was limited by contemporary western standards, and the state administration played a much stronger role. Political parties were also less developed than in western democracies and overloaded with nationalistic demands. Political participation was lower than in contemporary western states (Seton-Watson, 1945; Grzybowski, 1991). One aim of the Horthy regime was to restore a monarchical regime with traditional feudalistic values, institutions and authorities (Kitsch, 1970). Fascist parties were legal and were organised from 1932 onwards. The anti-fascist front was weak (Rothschild, 1988).

Independent Hungary, along with many other central and eastern European countries, experienced a polarization of political forces, starting with far left-wing Communist influence and then swinging to far right-wing conservative revenge, resulting in much unnecessary bloodshed. The counter-revolution between 1919 and 1941 was conservative in the sense that it re-introduced the monarchy, many feudalistic values, and other traditional institutions. It was also anti-liberal and chauvinistic. Apart from a short spell of fascist government (1944), the long period (1919-41) of the Horthy regime introduced some modest improvements in welfare and a certain degree of democratic political activity, though not including toleration of Communists.

From Democracy to Totalitarianism (1944-48)

After the war there was a short but important democratic period with a coalition government (1944-48). This period was very influential in the Hungarian "political memory" during the socialist era. Firstly, it was a proper multi-party system which accommodated parliamentary debates. Secondly, it was a period of an effective coalition of several parties. And finally, this period served as an important precedent to which the underground democratic opposition under socialism could refer as their historic example. Stalin, who did not direct the Hungarian Communists to seize power immediately because he hoped to win the locals over via elections, was shocked when in consequent elections (in 1945 and 1947) the Communists lost. This led them to change tactics and consequently to their taking political power by 1948.

By this time the international situation had also changed. The Cold War between East and West was under way, and Stalin became eager to strengthen his hold over east-central Europe for both military and political purposes (Grzybowski, 1991). He instructed his Hungarian comrades to change methods and finish the era of "parliamentary pirouetting". The democratic interlude in Hungary came to an end.

The Dark Ages of the Fifties, and 1956

What followed was the darkest period in the recent history of Hungary. It was fairly short historically speaking (lasting until 1953) but had tremendous consequences in political terms. The Communist regime under Mátyás Rákosi, "the best pupil of Stalin", showed its worst face to the people (Gáti, 1990). The reformist Imre Nagy, who became popular in 1953, represented an alternative to the hard-liner Rákosi. The power struggle between them created an unstable political situation. Moscow was also divided. Some supported Nagy and his reform ideas while the orthodox line bolstered Rákosi's side. Nagy's "New Course" policy had raised public expectations but hardline Communists were still in power all over the country. They included the prime minister and the party leadership.

The core of the opposition was made up of disillusioned intellectuals who had enthusiastically joined the Communist Party between 1945 and 1948 out of abhorrence at the deeds of the Nazis. The revelations of the early 1950s, however, transformed these people into the most determined opponents of the regime. Rejection of the Rákosi period was, however, the only common denominator among those who so enthusiastically joined the uprising. Both the wider population and even the smaller core of opposition stood for diverse views, which were often changing (Lomax, 1976). Nagy became the Prime Minister on 23 October 1956. His government comprised political parties of the post war democratic coalition period consisting of Smallholders, Social Democrats, National Peasant Party members, and Communists (including János Kádár, the post-1956 party leader). The Government of 1956 thus reflected the results of the elections of 1945 in its representation and expressed a desire for harmonious and friendly relations with the Soviet Union. Nagy demanded the removal of Soviet military forces from Hungary and withdrawal from the Warsaw Pact (the military organization of the Soviet bloc) (Grzybowski, 1991). He declared Hungary a neutral state and asked the United Nations to discuss the Hungarian case. People were hoping for intervention from the West but it was in no-one's interest at the time to provoke Soviet interests for the sake of saving Hungary and on 4 November 1956 Soviet troops overthrew the second, short-lived period of democracy in Hungary since World War II.

The Soviet leadership installed Kádár as the new leader of the country while Imre Nagy was made a political scapegoat. The 1956 uprising lasted only twelve days but became one of the best known events of Hungary's modern history. Its political consequences were also very significant. Although there was not much agreement among the various political factions in Hungary at the time, there was a growing rejection of Stalinist methods, resulting in the emergence of an unprecedented regime within the Soviet bloc.

The Liberal Kádár Regime

The new government firmly re-established itself. All the parties, except for the renamed Communist Party, the Hungarian Socialist Workers' Party, were dissolved and political pluralism ended. But the new socialist regime became very different from the one Rákosi and his partners had established ten years earlier.

There were a few basic points which Kádár would not alter. These were that (a) the political structure had to remain based on a strong one-party leadership, (b) Hungary had to remain a very close ally of the Soviet Union, (c) Soviet troops had to stay on the territory of the country, (d) Hungary had to remain a member of the Warsaw Pact and Comecon (the economic organization of the Soviet bloc), and (e) the supremacy of state ownership had to be kept. These were not principles of a democratic regime, but the Kádár regime proved that a relatively liberal political system and advanced economy could be introduced within such conditions.

In 1968 major economic reforms were introduced, known as the "New Economic Mechanism", developed by economists commissioned by the new government (Pető and Szakács, 1985). This aimed at introducing an economic system that could combine the positive aspects of central planning with the stimulation of the market and a price system which was closer to reality and, unlike the Soviet type of economy, created a buyers' market instead of a sellers' market.

Compulsory deliveries of farm products were abolished at the beginning of 1957, the control of agricultural prices was reduced and step by step they became market dictated. State intervention took the form of subsidies and taxation instead of planning regulations. Many agricultural products ended up with prices completely set by market forces. Dissolved agricultural farms were reorganised but without brutality. Collectivization by this time had become much more popular. Although heavy industry was kept going with huge state subsidies, mainly for military purposes, light industry and the service sector were rapidly developing.

The economic reform process concentrated on a new price system which allowed companies considerably more freedom and responsibility. Compulsory plans obliging enterprises to produce were completely abolished and central planning only operated at a macroeconomic level. Companies in the new system were regulated by taxes, wage guidelines, and so-called indirect regulators (Berend, 1990). They became profit oriented enterprises even if a large proportion of their profits was taxed. Nevertheless, the more profit they made, the more they could keep for themselves. This was a system unique in the socialist world.

While the economic reform was unprecedented there were very important principles which restricted the development of a full market-led system. One third of the enterprises, particularly the large ones, were heavily subsidised by the state. Another important policy was the prevention of unemployment. The political leadership tried to prevent inflation too, but without success (Berend, 1990).

The consequences of the radical and more or less consistent economic reforms were, however, significant. Hungarian agriculture became one of the few success stories in eastern Europe, and one third of the companies became successful, profit-making enterprises. This contributed to a growing proportion of the population (mainly younger and educated people) gaining experience in entrepreneurial skills. Living standards also grew to a relatively high level compared with neighbouring socialist societies. Every third family owned a car by the early 1980s and 60 per cent of new houses were privately owned. Private property and a limited private economic sector became part of Hungarian-style socialism. The infrastructure and the service sector also received much more attention than earlier. Increased investment made life easier than anywhere else in the Soviet bloc countries: roads and supermarkets were built, and public libraries and swimming pools opened all over the country (Bárány, 1990). The health service and the level of education improved radically and remained universal and free at the point of consumption, and there was a comprehensive and heavily subsidised child care system (Barr and Harrison, 1994).

In the middle of this overall development, however, there was a lack of change in the formal political sphere. The system remained loyal to its originally established principles and remained a one-party regime. The economic reforms only brought informal political changes in the regime. It thus became liberal and relatively tolerant of criticism. No-one was imprisoned for holding opposing political views, but no opposition political party was tolerated and social movements could only appear in the early 1980s. (The question of political opposition under socialism will be discussed in chapter 4).

The gradual changes under Kádár's liberal policies created a situation in Hungary from the late 1960s onwards where there was a continuous increase of living standards and relative tolerance in political terms. This allowed a very peaceful transition to the new, multi-party system in 1989-1990. Developments since the regime changed are the subject of the second part of this book.

Conclusion

I have shown in this chapter that Hungary became a semi-developed society prior to its socialist period. It became more developed than its neighbours to the East, but less developed than its western European counterparts. This "intermediate" situation was partly due to its location which resulted in frequent occupations by other empires. Hungary's more developed situation, as opposed to Russia, is due to its development more along western European lines when it was independent and, to a certain extent, when it was under Habsburg influence. Apart from that, Hungary did have a (limited) multi-party system between the two wars and two short spells of democracy since World War II. Hungary's relatively advanced democratic development, however, is also due to the process which occurred in its last period under socialism. As in the other central-eastern European societies, the majority of people strongly resented and resisted a Communist rule which, they felt, was imposed on them from the Soviet Union. From the 1960s there were major economic reforms and a relatively high degree of political tolerance which facilitated a smooth transition to a market economy and multi-party system after 1989.

By contrast, Hungary's democratic development lagged behind that of many western European societies for two major reasons. Firstly, because of the lack of political independence during centuries of foreign occupation and secondly, because of conservative internal forces. Feudalistic institutions were conserved in Hungary long after they had lost political power in western Europe, and right wing, chauvinistic values were maintained up to the end of World War II (to re-emerge again after 1989). In the next chapter I will examine Russia's democratic development.

Notes

1. The invasions of the Mongols (thirteenth century), Turks (1526-1772) and Habsburgs (1773-1918).

2. The Reformation reached Hungary very early and won over the majority of the population at the time. Instead of Latin, widespread usage of the native language was encouraged and the Bible was translated into Hungarian as early as 1590. From the end of the sixteenth century, however, the Habsburgs re-introduced Catholicism wherever they could within Hungary. This led to a religious division which remains even today: western Hungary is mainly Roman Catholic and the part to the east of the Danube is mainly Protestant (Calvinist and Lutheran).

3. There were 2.2 million Romanians, 1.6 million Slovaks, 1.3 million Croats, 1.2 million Germans, 0.8 million Serbs and 0.4 million Ruthenians in 1840 (Deák, 1990).

3

The Historical Absence of a Democratic Tradition: the Case of Russia

Introduction

The previous chapter demonstrated the strong continuities between Hungary today and its historical past. The present political turmoil and the politically underdeveloped environment in Russia also has its roots in the past. Everyone is aware of the Stalinist period which eradicated all trace of democracy in the Soviet Union, but the roots of the lack of development of democratic institutions in Russia go back a lot further. Russia's historic development throughout several centuries created a very specific political tradition and culture which was substantially different from a "western" or "central-eastern" European model. It will be demonstrated in this chapter that Russia has never experienced a period when democracy and parliamentarism could have served as a basis for current democratic development. The more we look at the pre-Communist period the more the continuity with the present becomes clear.

I will argue that the lack of democracy in Russia today is linked to a tradition of authoritarian rule, the origins of which lie in the autocratic style of rule which created a different cultural tradition resulting in a different understanding of the notion of democratization in Russia today. This is linked to the collectivist style of Byzantine Christianity and the strong historical link between state and church. It is also linked to the late abolition of serfdom and the consequent delayed economic development, the weak development of the middle class, the bureaucratized administration and the centralized system of government which are closely connected to the development of the Soviet-type socialism. All these led to a different understanding of the notion of democratization in Russia.

The Beginnings

Russian civilization is over a thousand years old. Some argue (Thompson, 1990), that Russia's first period (the first state was the Kieven Rus', c. AD 800-1200) already had a different pattern of peasant-landowner relationship from that characteristic of western Europe at the time, a pattern of obligation which tied essentially free peasants to a status approaching slavery. Riasanovsky (1984), however, claims that the Kieven Principality was based on democracy, with well developed local institutions in the towns led by the "duma" or council of boyars, an immediate predecessor of a parliament, and regular town meetings, an assembly of freemen. If Riasanovsky is right then the Kieven Rus' was the *only* possible example of an embryonic democratic institution in Russian history. The adoption of Byzantine Christianity[1], however, enforced a different path of development. One very important characteristic of Orthodox Christianity, in sharp contrast to western Christianity, was its deep concern with the collective spirit of the whole congregation, as opposed to the individual. Collectivism and the collective sense of community was the basis of the development of the Russian peasant community, the so-called "obschina" (Riasanovsky, 1984). The introduction of Christianity, a thousand years ago, affected the legal and the political system, education and culture as well as the religious aspects of life in the developing Russian society: written language emerged which raised the general level of literacy, education and literature.

With the emergence in the sixteenth century of the Moscow Principality as the dominant power and unifier of the Russian lands (Kluchevsky, 1968), a concentration of power took place ensuring that any other political concept that coexisted at the time (like the republican, democratic government of Novgorod, and the boyar or aristocratic domination of Lithuania) was subordinated to the autocratic system of Moscow (Andreyev, 1976). The first unification, provided by the Moscow state, offered to the Russians a reasonable degree of economic recovery and peace against the surrounding enemy but it established an absolutist, centralized rule in return. The power of the tsar to command obedience was based on the Byzantine tradition of the Orthodox Church, which strongly supported autocracy and hence reinforced the authoritarian style of leadership of the Russian tsars for centuries ahead (Walsh, 1963).

Lagging Behind with Modernity

The subsequent centuries after the period of unification followed this "non-western" path. While in western Europe classes and institutions, competitive with the king, were growing stronger, an opposite process was developing in Russia. Tsars had regularly and successfully undercut the power of boyars and any other groups which might have challenged their power, and no alternative basis to organize political opposition was present (Sumner, 1943).

Administration and bureaucracy also continued to expand, thereby ensuring that the tsars' orders were followed. The role of the tsar became highly personalized and governments played a small and inefficient role in the ruling of the state. At the time when in some parts of western Europe the accumulation of wealth was laying the basis for the Industrial Revolution, the Russian economy developed only very slowly and started to lag behind western technology, science and thought (Thompson, 1990). The new farming techniques and technology generated an agricultural surplus in western Europe leading to profitable local, regional and international trade and the enrichment of merchants especially in urban areas. Meanwhile in Russia the economy remained predominantly agricultural and agriculture remained at a low subsistence level. Technology did not develop and the level of productivity remained low. There was little commercial surplus and even that was concentrated in the hands of a relatively small group of rich merchants. Trade, technology and urbanization, the most important forces of modernization in western societies, grew very slowly in Russia (Riasanovsky, 1984). The western Europeans were in the middle of overseas discoveries and expanding trades beyond previous boundaries aimed at new continents, while Russia was involved in wasteful bloody wars aiming at gaining more territory but not improving her economy (Thompson, 1990).

The lack of modernization became Russia's main problem from the sixteenth century onwards and has remained one of the biggest obstacles up to the twentieth century. Class divisions hardened, widening the gap between a small elite and the bulk of the society. A particularly large proportion of peasants became enserfed just at the time when serfdom was disappearing in western Europe (60-70 per cent of the whole population). Serfs were practically slaves. They became personal chattels of the landlord, and could be sold apart from the land and separated from their families. Landlords not only required labour but exercised judicial and administrative control over their serfs, a system which remained a main pillar of Russian society lasting up to 1861.

There were some attempts to modernize the traditional and backward country (Bogoslovskij, 1963) but these were usually met by enormous resistance amongst the ruling classes who successfully opposed any alteration of the traditional system. There was also an absence of any other strong social force pursuing modernization: the merchants were weak, no potential entrepreneurial class was developing and the nobles were divided (Spector, 1965). In addition, education, including higher education, was almost nonexistent; corporate bodies, such as guilds and other burgher associations, which in western Europe played such an important part in the growing political role of the bourgeoisie by resisting authoritarianism, were also absent (Oliva, 1969). Any attempt to move Russia into a more secular society, in which people acted as citizens with more individual rights and demands, rather than as part of a secluded "obschina", was in vain (Rozhdenstvenskij, 1963; Spector, 1965; Massie, 1981). Fundamental problems, like the privileges of the nobility, the corrupt bureaucracy and lack of education, were not addressed. Serfdom, which became socially and economically the most anachronistic and ultimately the chief obstacle in Russia, holding the country back from the kind of social, economic and political development contemporary western European societies experienced, thus had serious consequences (Lentin, 1973; Jones, 1973b). As a result of the lack of political pressure for reform, the regime failed to be modernized (Soloveytchik, 1945).

The Nineteenth Century

By the early nineteenth century Russia had become a huge empire of 41 million people[2] containing many ethnic groups with a strong official policy of Russification (Seton-Watson, 1967). The overwhelming majority (99 per cent) of the population lived in rural areas and only two million in towns and cities which were run by an extensive and extremely inefficient bureaucracy headed by an authoritarian government and an autocratic tsar. Civil liberties were limited, and censorship and repression were extensive (Thompson, 1990). The overwhelming majority of the population was made up of ill-educated, privately owned serfs and state peasants (almost forty million). The serf economy continued to conserve a system with few incentives to increase production and surplus. There was no interest in investing in agriculture or accumulating capital for large-scale industry. Traditional practices, producing only a moderate surplus, provided the proportionally very small group of noble landowners with enough income for their luxurious lifestyle (Dukes, 1990) and serfdom permitted them to control the bulk of the "enslaved" population who lived at a bare existence level. A ruthless tax collection system placed heavy burdens on the peasants in particular and they provided human resources for the ever important army as well. This backward regime with its hierarchical interest, helping the few at the expense of the majority, maintained a huge and diverse empire in a hugely underdeveloped state falling centuries behind its western counterparts (Yaney, 1973).

Delayed Modernization

In the period after the emancipation of the serfs in 1861 the structure of

Russian society still followed a strict hierarchy (Troyat, 1961). The 1897 census reveals that only 1 per cent of the population constituted the hereditary noble class (Seton-Watson, 1967). These were landowners and those impoverished nobles who occupied the highest ranks of the armed forces and the civil bureaucracy, which doubled in numbers between 1880 and 1914 (Manning, 1982). There was a tiny (1 per cent) urban merchant (business) class, urban professionals (0.3 per cent) and 5 per cent industrial working class, concentrating in a few industrial areas. The recorded number of peasants, on the other hand, was 99 million people (85 per cent of the total population of 116 million) (Seton-Watson, 1967; Acton, 1992). The gulf between the ruling class and the "ruled" was thus enormous in Russia at the turn of the century.

However, at this time Russia underwent a much delayed transformation (Field, 1976; Keep, 1976). Industry grew by drawing heavily on European capital for investment, using Russia's rich natural resources, the vast pool of available peasant labour[3], and government support in the form of subsidies, tariffs, large government orders and outright ownership of key sectors such as railroads (Robinson, 1970). However, the continuing pre-eminence of the state was highly significant in this process. As a force, independent of the government, capitalism never really got off the ground in Russia. Many features of the "planned economy" introduced after 1917 had pre-revolutionary antecedents (Keep, 1976; Bater, 1987). The developing industry concentrated in a few major cities and regions (like Moscow, St. Petersburg, the Urals and the Ukraine). It was dominated by the state, on the one hand, and foreign investment, on the other (Pokrovsky, 1970).

The social consequences of this rapid development were serious (Tsagalov, 1970). Although foreign investors played a prominent role, a Russian capitalist class did not develop with the economic strength and political power to force through badly needed political changes. The number of workers shot up to 2.25 million, which was a great improvement, but they still represented a relatively small proportion of the whole population. Cities expanded in number and size but their living conditions were similar to those of western European cities centuries earlier (Bater, 1987) and agriculture remained backward and inefficient (Perrie, 1989).

While western political parties developed with the concept of representative institutions, in Russia they grew up in the absence of such institutions (Haimson, 1970). Prior to the 1830s, groups or individuals who felt oppressed by the authority, saw the political solution as either to change "the person of the supreme arbiter" (the tsar) or the dynasty while leaving the existing framework of political and social organization intact, or to revolt against the political and social order, a wholesale rejection without any conception of an alternative. Later (from the 1840s) utopian socialist views started to spread among the more western-oriented Russian intelligentsia many of whom believed that Russia could bypass the capitalist stage of economic development and thus pass directly from semi-feudal conditions to socialism. This belief was based on the assumption that the Russian peasant commune was an instinctively socialistic, egalitarian and democratic institution which would enable Russian peasants to build socialist collectivism[4] (Offord, 1986). The other important political trend was terrorism. One of the aims was to stir up the peasants, and arouse them to protest. Terrorism was considered the best weapon of agitation to make the peasants more receptive to socialist propaganda. It was only a decade later in the 1890s that revolutionaries became increasingly aware of the receptivity of the urban workers to socialist ideas and many of the intelligentsia radicalized and adopted Marxist ideas (Crankshaw, 1978; Acton, 1992).

The two major wings of the revolutionary trend became the Social Democrats, the party which was orientated towards the urban proletariat, and the Party of Socialist Revolutionaries which advocated the need for an agrarian revolution based on the peasantry. At the turn of the century the most important political development was a wide ranging coalition, an imposing front against autocracy, ranging from zemstvo constitutionalists on the right, to the professional unions in the political centre, and to the Social Democrats and Socialist Revolutionaries on the left, agreeable to sticking together against absolutism, as a common immediate political objective (Haimson, 1970).

In an economic and social sense Russia became a relatively rapidly transforming society by the turn of the twentieth century, but the tsarist political system still ensured an antiquated conservative political regime, resisting concessions and blocking evolutionary attempts at change (Pearson, 1989; Lincoln, 1990).

There were three major revolutions in Russia within twelve years: one in 1905 and two in 1917. Prior to that there were only isolated and brutally suppressed uprisings. Thus the revolution of 1905 was the first movement in the history of Russia to have a broad social basis, a largely spontaneous popular rejection of the tsarist regime, and as some have argued (Sablinsky, 1976), the first major event to set the patterns and unleash the forces that succeeded in 1917. In 1905 people were calling for radical reforms, demanding equal status for the peasantry; civil freedom and freedom of the press, assembly and speech; a representative legislative assembly and independent associations for professionals. As a result, a new constitution was introduced, which gave all Russian males a right to vote. A main assembly of elected representatives (*Duma*) was set up (in 1906) but there was a weighting system, in which the vote of landowners and large urban property holders had 250 times more weight than that of a peasant's and 125 times more than a worker's (Seton-Watson, 1967).

Newly legalised trade unions were set up and the workers started to organize coordinating committees of elected representatives of the factories to direct and supervise strike action. These committees were modelled on the traditional communal assemblies of villages, and became known as Councils of Workers' Deputies, and later Soviets. Originally they were spontaneously growing grassroots organizations, but they soon became platforms for revolutionary agitation against factory owners and the government.

During the pre-World War I industrial boom, Russia's industrial output grew and its labour force reached 3.5 million workers, many of whom became radical as a result of the unresolved social tensions and active political propaganda activity (Pushkarev, 1963; Swain, 1983). The propaganda activities of the revolutionary intelligentsia also grew sharply, particularly in urban areas. However, by 1914 the continuing resistance of the tsar and the conservative government to implementing reforms, the structure of the state bureaucracy and the lack of an established mechanism for further political innovation had brought the country to boiling point, uniting the different sections of the opposition and leading to the demand for constitutional democracy.

The most radical segment of the educated elite, however, looked beyond the idea of a constitutional democracy (Acton, 1992). Two major radical parties became important by around 1910, the Marxist-inspired Russian Social Democratic Workers' Party and the peasant oriented Socialist Revolutionary Party.

The Social Democrats were split into two factions. The *Mensheviks* wanted a broad legal party embracing wide sections of the workers and advocated cooperation with the liberal representatives in the Duma. The *Bolsheviks,* however, favoured a centralized party of professional revolutionaries and underground activities, and saw peasants only as potential allies. By 1914 the Bolsheviks became the most influential amongst industrial workers in major cities (Acton, 1992). Although the outbreak of World War I initially brought a certain degree of national unity and patriotic fervour, by 1916-17 the mounting number of deaths and casualties lowered morale among the 15 million predominantly peasant conscripts and the civilian population leading to growing discontent, demonstrations and strikes (Fischer, 1970).

Thus, while the 1905 Revolution opened up new opportunities in the democratic modernization process in Russia it failed to deliver most of its promises which, combined with the military defeat, created a growing social tension leading to the revolutions in 1917. The February Revolution abolished the centuries old tsardom in Russia and the discredited autocratic regime collapsed with surprisingly little resistance (Katkov, 1967).

By 1917 the major political debates were between the Mensheviks and Bolsheviks. The Mensheviks, insisting on their interpretation of Marxism, argued that Russia should follow the capitalist phase of development before they could undergo a proletarian revolution and enter the final stage of history, socialism. The Bolsheviks, on the other hand, had by April 1917 overtly abandoned the idea of evolutionary development and were openly aiming for full political power (Fitzpatrick, 1979). The Bolshevik Revolution was more of a political achievement than a military victory (Liebman, 1970). The real battle was the exhausting civil war which lasted for several years (1918-21).

After the October 1917 victory the first act of the newly elected Council of People's Commissars was to announce a national election (13 November 1917) the results of which were as follows: Social Revolutionaries[5]: 58 per cent, Bolsheviks: 25 per cent, Bourgeois parties (Cadets, etc.): 13 per cent, Mensheviks: 4 per cent. In other words the majority of the Russian population in October 1917 was Socialist oriented, not Bolshevik (Mawdsley, 1987).

In interpreting the October 1917 revolution many have argued (Rosenberg, 1987), that in 1917 in Russia the concepts of democracy, civil liberties and rule of law scarcely existed. Workers and peasants were closer to the notion of social dominance (*vlast*), an "almighty" power, and lacked of any historic experience of parliamentarian, gradual solutions to political and social problems. In Russia, society both at the top and the bottom, had never known tolerance and the rapid political upheaval allowed people to "wrap a cloak of theory around the violence" (Ferro, 1985, p.23).

The major question for this chapter is how strong the impact of the lack of democratic traditions prior to 1917 was and whether the stamp of the forthcoming Soviet regime was made in October or later. In other words whether the features of one-party rule and terror were inherent in the Bolshevik power. Some (Koenker, 1987) argue that they were not inherent Bolshevik features, but were Stalinist ones. Others (Rosenberg, 1987) support the view that the Bolsheviks introduced terror and dictatorship as soon as they came in power.

By and large, people in Russia were politically inexperienced and had not had the chance to develop trust in a system of evolutionary progress based on parliamentarism. Thus the Bolshevik revolution in October 1917 was not based on wide social agreement, as portrayed by mainly Soviet historians. It was a successful coup by a narrow group of highly discontented and well organized skilled workers mainly in urban industrial areas led by politically active "professional revolutionaries". But, soon after, many workers, who had previously sympathized with the Bolsheviks in their radical demands, became critical about them when their hopes were not fulfilled immediately. Their lack of political experience led them firstly, to believe that promises and hopes of rapid changes could be fulfilled and secondly, made them discontented and impatient within a short space of time. It was the coming period of Civil War which decided the final outcome of the Bolshevik take-over.

The Soviet Period

The USSR (Union of Soviet Socialist Republics) officially came into existence only in 1922 but in January 1918 Lenin declared the new regime to be the *Soviet Republic of Workers, Soldiers and Peasants* (Liebman, 1970). As Harding (1984) argued, "Lenin's conception of Soviet democracy clearly went a good deal further than Marx's sketch of the Commune" (p.20), because Lenin was over-optimistic in his approach when he predicted that democratic control of the economy and planning would be achieved on the basis of "free and conscious relationship of man to man ... and socialist ownership ... [which] would extend itself through its technical superiority and natural attractiveness" (p.21). Forced to solve a host of less theoretical problems the Bolsheviks faced a hefty dose of pragmatism.

One of these problems was that the Bolsheviks were a clear minority in the Constituent Assembly after the November 1917 general elections.[6] In order to solve this they simply disbanded the Constituent Assembly by January 1918 (with a decree which was opposed even by a minority of the Bolshevik delegates) (Liebman, 1970). This decree brought to an end the very brief parliamentary episode of the history of Soviet Russia and by 1920 all the major leaders (Lenin, Trotsky and Bukharin) accepted the idea of "the dictatorship of the proletariat by the means of the dictatorship of the Party" (Hill, 1989, p.168). It was, however, still the Bolsheviks' view at the time that nationalization and centralization of *all* means of production and distribution was not in the best interest of the economy, and thus these were measures not implemented immediately.[7]

Between 1918 and 1921 the country suffered from the chaos of civil war and foreign intervention. To make the best use of scarce resources the Bolshevik leadership tried to centralize supply and output and nationalization was speeded up in a wide range of industries (e.g. mines, metallurgy, transport) (Wood, 1979). This was a measure to step up the production of arms rather than a decision based on theoretical or political principles (even if it was in line with original Communist ideas). The method itself became known as "War Communism". Trade and industry became a state monopoly and in industry large and medium sized concerns, serving the national market, were brought under central or provincial state control.

In 1917 the Russian Empire was one of the largest on earth with a mainly rural population. During the civil war the country had to face two major problems. Firstly, that the majority of people supported the Socialists and were against the Bolshevik power. The Socialists were, however, badly organized and attracted the mainly rural, mostly illiterate part of the population. The Bolshevik strongholds were the few major cities. The second major problem of Soviet Russia was the vast proportion of ethnic minorities.

The ethnic minorities (half of the total population of 160 million) were a significant factor in the outcome of the Civil War. The empire's minorities were very varied in 1917. Although the Russians were the largest group, they constituted only half the population. Apart from the three other largest minority groups[8], the rest of the numerous minority nations were not even Slavs.[9] One of the problems stemmed from the fact that the educational, cultural and political level of the minorities was even lower than that of the Russians (Mawdsley, 1987). The socialist and the original Bolshevik propaganda appealed to many groups among them and, although the Bolshevik Revolution was dominated by Russians, it was fought on the basis of "class struggle". This appealed to many ethnic minorities even more than goals based on nationaliztic principles (Suny, 1987). Russian society was also stratified along religious/ethnic lines. The minorities' contribution to the Civil War therefore was not negligible.

The bloody Civil War, which lasted for three years, was unique in many senses. Firstly, it was not simply an internal war, but a war with substantial intervention by the Allied forces supporting the Whites against the Bolsheviks. Secondly, it was a class war (Serge, 1972; Manicas, 1989), and finally, it was unique because nationalizm played an important role in it (Schapiro, 1984; Lewin, 1985; Wiliams, 1987).

The Introduction of the New Economic Policy (NEP)

The introduction of the NEP was a response to the inadequacy of War Communism in economic terms, and the outcome of the many political debates provoked by social and economic failure. The NEP was understood by many in the West, as well as within the Soviet society, as a retreat from planning and communism in general (Lane, 1996).

Some (such as Trotsky, Zinoviev and Kamenev) argued that the NEP was a clear diversion from Communist ideas and led the "leftist" opposition within the party (Kochan, 1983). Others, Bucharin, for example, valued it as a clever compromise. Using Lenin's well known phrase, it was a policy of "one step backward, two steps forward" but it was "prudent pragmatism" (Lincoln, 1989). The NEP certainly introduced many new economic policies which the official socialist propaganda had originally firmly resented. The desperate economic situation in which the long lasting Civil War wrecked the country forced the Communist leadership, Lenin above all, to compromise on their ideological principles and allow some "capitalist" features to come into action to stimulate the economy even if reluctantly and as a temporary necessity (Schapiro, 1984). Under the NEP some elements of the free market and private ownership were reintroduced, combined with substantially less government control in the labour market and trade than before.[10]

These relatively limited changes led to an unexpected success. The semi-nationalized and semi-private structure of the NEP proved that even smaller concessions worked exceedingly well compared with the fully Sovietised system of War Communism. The extensive starvation of 1921-22 had ended and production had recovered to reach its pre World War I level. Productivity grew, trade revived and shops reopened. Job seekers poured into the cities from the countryside but wages fell below the 1914 level while prices rose as a result of an excess of demand over supply. Still, a sense of normalcy and hope returned after the chaos and famine of the Civil War and War Communism (Wood, 1979; Nove, 1989; Littlejohn, 1984). It was, however, only at the level of the economy that some relaxation was tolerated (Mackenzie, 1977; Lincoln, 1989). The Communist Party retained full political control throughout the whole period of NEP (1921-28).

In sum, the Bolshevik leaders, many of whom had previously lived in the West in exile, recognized the backwardness of the Russian state at the turn of the century. Unlike the Menshevik wing of the left, however, who believed in evolutionary development via parliamentary means, the Bolsheviks, although divided on many issues, shared the view that only rapid modernization would help to bring Russia into the twentieth century. As the majority of people supported the Socialists in 1917 rather than the Bolsheviks, they had a difficult task on their hands in winning over the resistance against a Communist power. Apart from the political (and military) tasks, it was the economy which created considerable problems for the young Soviet state. The failure of War Communism led to the temporary "relaxation" of Bolshevik principles with the introduction of the NEP. Although it was successful in economic terms, it contradicted Marxist postulates. This created doubts in many Bolshevik leaders' minds and paved the way for a return to the original Marxist-Leninist principles.

The Stalinist Period

Apart from the debates concerning economic policies and international affairs, the question of succession surfaced in 1922 when Lenin became ill. The natural successor of Lenin was Trotsky, the second person in the party. Stalin was neither a strong theoretician nor a particularly popular person, but he was a very obedient administrator (Lincoln, 1989). When Lenin became ill Stalin skilfully outmanoeuvred his rivals. First, he allied withTrotsky, then with those who were considered to be at the "right wing" of the Bolsheviks, among them Bukharin, against the so-called "left wing", such as Trotsky. Finally, he suppressed both rival groups by turning them against each other. This way he managed to achieve supreme authority within the party and consequently in the whole Soviet Union (Riasanovsky, 1977; Thompson, 1990). "Stalin was a secretive man and his published works and speeches give us less insight into his real thoughts than is usual with politicians. Indeed he lied on a prodigious scale." (Nove, 1984, p.34). Consequently it cannot be known when precisely he made up his mind to turn against the NEP policy. According to Nove (1981) and Carr (1979) Stalin was probably always against the NEP and regarded it as a temporary compromise, preferring ruthless "strong-arm" methods.

The questions here are: what was the relationship between Bolshevism and Stalinism; and whether Stalinism was inevitable. Opinions and arguments on these questions are sharply divided and can be broadly grouped in two. Firstly, the majority of western Sovietologists (like Armstrong (1961), Daniels (1962), Treadgold (1964), Brzezinski (1966), Tucker (1971), McNeal (1975) and Fainsod (1979)), identified Stalinism with Bolshevism and totalitarianism. This was the case especially during the cold war period but some argued this way even later. Such writers often saw the Stalinist regime as an inevitable development in the Soviet system right from October 1917. Part of this group stressed that Bolshevism between 1917 and 1928 already contained the "germs" of Stalinism and that it was Leninism, rather than Bolshevism, which was nascently "Stalinist".

The arguments of the other group, however, question the consensus on continuity and totalitarianism (Lewin, 1968; Fitzpatrick, 1979; Cohen, 1985). It is also emphasised by some within this group (Burnham, 1962; Djilas, 1957; Kolakowski, 1977) that it was only under Stalin's leadership that the regime developed a bureaucratic order and terror. The main difference between Soviet authoritarianism before and after Stalin, it was argued, was not simply nationalizm, bureaucratization, the absence of democracy, censorship, police repression, because these are characteristics of many societies. Rather, Stalinism was an excess, an extraordinary extremism in each, a holocaust by terror that victimised tens of millions of people for twenty-five years (Cohen, 1986).

In fact, as Hannah Arendt (1973) pointed out, Stalin not only became a one-party dictator in one society, he created a totalitarian movement all over the world. He liquidated all factions within the Communist movement by abolishing internal party democracy and transforming national Communist parties into Moscow-directed branches of the Comintern. Nazi totalitarianism, argued Arendt, started with a mass organization and was gradually transformed into elite formations which dominated them. The Bolsheviks, on the other hand, started with the elite groups and organized the masses afterwards. The common element in the two regimes was that they were both based on the secret police, the secret society of conspirators (Arendt, 1973).

It was also argued (Kochan, 1983) that Stalin, and the Communist party, carried out a second "revolution" in Soviet society from 1928-29. This caused a major turmoil in the country with enormous suffering. The aim was to force through rapid industrialization and collectivization to create an economically and militarily powerful and strong Soviet Union. This policy was partly based on a simplified version of the Marxist ideology, namely that a society during its development had to become highly industrialized and must have an overwhelming dominance of the working class. On the other hand, it was a response to the feared renewal of foreign military intervention.

To carry out rapid modernization based on a "scientific" approach, the State Planning Commission, the so-called Gosplan, was given the task of drawing up three five-year plans between 1928 and 1941. One aspect of the state planning system was collectivization, forcing all the peasants to transfer their newly distributed lands into collective farms. The collectivization process, however, turned out to be a social disaster at the time of its introduction and created what remained a very inefficient system of agricultural production throughout the Soviet era (Kochan, 1983). The peasants felt robbed of their recently distributed lands and agricultural output dropped substantially. As a result, by 1933 the outbreak of famine in many parts of the Soviet Union claimed the lives of millions. Stalinism did not overcome the productivity crisis in agriculture but rather assimilated it. It created an "everlasting imminence of crisis. Stalinism carried the relations which engendered the crisis to their ultimate limits" (Campeanu, 1988, p.123). The monopoly situation transformed the periodic crises of agriculture into a chronic condition.

Industrialization accomplished more, though at a price of countless sacrifices. In the field of heavy industry, if we discount environmental and human damage, substantial success was achieved. The output of heavy industry grew at 12-14 per cent a year during the Stalinist period[11] (D'Encausse, 1981) and the urban population doubled (Thompson, 1990). The main target was to achieve a large *quantity* of output. Quality became a secondary aspect and was largely ignored by Soviet planners. Light industry, most of the infrastructure and the service sector were consistently ignored. The central planning system remained a very important feature throughout the whole Soviet period. It provided, or so it seemed, an opportunity to keep everything under tight control and hence ensured constant development for the entire economy.

The ideology of the mono-organizational system, introduced by Stalin, was wholly monopolistic in its claims (Rigby, 1990). People in the Stalinist regime were forced to go through a legitimising process: they were forced to accept certain values, attitudes, and behaviour-patterns based on an ideology claimed to be Marxist-Leninist. The Soviet Union reduced itself to achieving a single, all-embracing goal, the construction of Communism—an impossible task which naturally led to failure (Rigby, 1990). The main characteristics of the Stalinist period were authoritarian leadership, control over every segment of life possible, suspicion and resort to terror (Mackenzie, 1977). The Bolsheviks were purging the very class whose hegemony the regime was to embody. The blacklists, the punitive labour platoons and the concentration camps, which were supposed to have been invented for the so-called "class enemy" and political adversaries, were rapidly filled with ordinary workers and peasants, whose "crime" was that they refused to work under the inhuman circumstances they were forced into. The naked terror antagonized ordinary people and led to a widespread discontent. Political and economic change was inevitable.

The Period of Destalinization

When Stalin died in 1953 he left behind a devastated country. The Soviet Union was politically demoralised and economically ruined. This was partly the result of the Stalinist leadership and partly because of the devastation of World War II. Twenty million people died fighting against the Nazis and another twenty to thirty million in Soviet labour camps. The new party leadership recognized the danger of the one-person authority and opted for a collective leadership (D'Encausse, 1981). Having recognized all the damage, Khrushchev announced his "de-Stalinization" campaign.[13] While acknowledging some of Stalin's achievements regarding industrialization, Khrushchev bravely and unexpectedly denounced Stalin for his political style, the so-called "personality cult", his paranoid suspicions, the purges and labour camps, Stalin's failure to recognize the Nazi danger which caused unnecessary war devastation, and for the economic depression in the post-war period (Nove, 1989).

Under Khrushchev the political atmosphere eased considerably. Political opposition, which was more or less eradicated under Stalin, developed again (see chapter 2). As radical and innovative as these political changes were, no consideration was given to fundamental principles of the Soviet political regime, such as the one-party system. It was argued (Nove, 1989) that by the 1950s for most Soviet citizens the one-party regime had become normal, even natural. "What," said a Muscovite when a foreigner suggested a multi-party system, "several parties? Isn't one bad enough?" (Nove, p.122). Production and productivity in industry and agriculture increased in the new "reform" period, consumer goods and services were given higher priority and the general political atmosphere improved considerably. The housing situation and living standards also improved radically. Khrushchev was popular among ordinary people, intellectuals in particular, for the relaxed atmosphere he achieved, but his fellow leaders were suspicious of his approach. When the rate of industrial growth (the main yardstick of success used by the Communist leadership) dropped, Khrushchev found himself under heavy attack by some of his colleagues, especially the most conservative hard-line core within the party and on 15 October 1964 he was axed from the party leadership (Breslauer, 1980).

Thus the changes under Khrushchev were enormous. He emptied concentration camps, poured money into agriculture and some into consumption and services, and granted a certain degree of autonomy for businesses. He even allowed some germs of market elements to develop, as in the 1920s, by reference to the NEP ideas rather than to contemporary capitalist examples. Khrushchev was a Bolshevik, as Besancon (1978) argued, who wanted to introduce socialism "with butter". Khrushchev's attempts failed, however, because the Soviet regime was politically and ideologically driven.

The Last Soviet Hardline Period

Brezhnev, the new leader, brought back bureaucratic stability and some of elements reminiscent of the Stalinist regime, an "administrative command system", in Mary McAuley's (1992) phrase. Brezhnev, like Stalin, firmly believed in a centralizing power both politically and in the economy and brutally cracked down on any political dissent or cultural creativity. The relatively liberal atmosphere of the Khrushchev era ended dramatically when Brezhnev resumed political power.

The economy did not develop sufficiently under Brezhnev's leadership but science and education were encouraged, if only in a very traditional style, and always pledging full loyalty towards the state and the dominant ideology. The arms race continued to be a very important element in the economy, as a result of which living standards improved only to a very modest extent (McAuley, 1992). Keeping the whole society in a stable but traditional style seemed to be the most important element of the policy. The strong feeling of stagnation, however, frustrated many people, some in positions of leadership and it is mistaken to believe that Soviet society was completely frozen in the Brezhnev era.

Public opinion was developing independently of the opinion of the Kremlin leaders, as will be discussed in chapter 4. As soon as the threat of terror started to ease (in the 1970s) and later disappear (in the 1980s), social and political discontent was expressed more and more frequently (Mandel, 1989). Among those who expressed independent opinion were well known scientists, writers and artists. Despite the censorship, critical articles appeared from time to time in well known periodicals such as the *Kommunist* and *Novij Mir*. Young technocrats were also critical of the "bureaucratic despotism" of extensive industrialization. Young people often expressed rebellious views concerning the cultural isolation from the west, particularly in pop music, modern art and jazz (Mandel, 1989). A number of activists went as far as to organize protest groups (see chapter 4).

The Brezhnevian official doctrine, however, was out of step of these changes. It reintroduced a regime with a rigid, outdated view, insisting on Marxist-Leninist principles. Only two social classes were, for example, recognized, the working class and collective farmers. The difference between agricultural workers and industrial workers was seen as a difference between urban and rural. "Urban lifestyle" was considered to be more "advanced". Another important distinction describing the Soviet class system separated blue collar from white collar occupations. The intelligentsia was defined as a distinctive stratum with special skills, obtained by higher education, whose task was to serve the "leading" working class, and its allies, the peasants. Nevertheless, as education was given a strong emphasis by the regime, the intelligentsia was expected and did become the most rapidly growing group under the Soviet system (Kelly, 1986). Within this basic structure finer distinctions were also made between skilled, semi-skilled and unskilled industrial and agricultural workers. In the late 1970s the industrial labour force still consisted of a large proportion (up to 50 per cent) of unskilled and semiskilled workers involved in technologically fairly primitive work and only 17 per cent highly skilled workers. As a result of rapid urbanization the agricultural labour force was rapidly declining (from 58 per cent to 21 per cent between 1956 and 1977), but comprised mainly unskilled labourers (up to 80 per cent). The service sector barely existed (5 per cent). The category of white collar workers absorbed most of the rapidly increasing number of higher educated people (Hough and Fainsod, 1979; Kelly, 1986).

Thus the Soviet Union under Brezhnev's eighteen-year rule returned in many ways to a rigid, overcentralized command system with very limited tolerance towards deviation from the official ideology. It provided a strong sense of stability, especially in economic terms, but little scope for innovation. The stagnation maintained living standards but did not allow improvement which generated a discontent leading to the pressure for yet another, now more radical, change.

The Last Phase of the Soviet Period

Mikhail Gorbachev was elected as the last general secretary of the Soviet Communist party in March 1985. Recognising the stagnation and growing discontent, Gorbachev set in train the most radical reforms in the history of the Soviet Union. Gorbachev, however, never intended to bring about any of the fundamental changes his efforts led to by the end of his period. He, and his team, only intended to raise living standards and end political rigidity. In that respect he was wholly successful: the Soviet Union had not gone through such fundamental upheavals since Stalin's Revolution of 1928-29 (Dawisha, 1990). Gorbachev's motives were obviously complex. As a member of the younger, better educated, generation of communists, he was not prepared to continue with the paternalistic domination of a narrow-minded oligarchic leadership. There was also growing international pressure over the lack of civil rights in the country (Hosking, 1990).

Gorbachev's aims were largely pragmatic. On the one hand he was aiming at rationalizing the economy in order to achieve higher production, better quality goods, profitable companies and a kind of market mechanism within the Soviet framework which would stimulate the economy sufficiently while leaving the socialist ideological structure unchanged. If this sounds contradictory, so it was in the minds of Gorbachev and his associates (White, 1991). The other important element of Gorbachev's reform was *Glasnost*, the political reform. Gorbachev's political pluralism allowed the Soviet Union to undergo an unknown experience in political terms. Although political reforms were second in Gorbachev's priority to economic growth they actually played a more important role consequently (White, 1991).

It became easier to change the political system radically while the economic problems remained largely unsolved or were worsening. If the central objective of Gorbachev's reform strategy was to recover the economy's growth dynamic, it was clear that it fell far short by a long way. In fact, the late 1980s and early 1990s were periods of deepening economic crisis. But Gorbachev, restricted by his Communist beliefs, could not lead the reforms far enough: he could not accept the idea of a multi-party system in the Soviet Union. His idea of "political openness" remained based on the primacy of the Communist Party in the Soviet Union. When, from January 1989 the question of a multi-party system was frequently raised by many, it was firmly rejected by Gorbachev. He and his supporters stuck to the view that democracy could be achieved within the existing political framework. The existence of a multiplicity of organizations and associations expressing a wide range of social interests while improving democratic processes within the one-party regime was a good enough guarantee for him.

Gorbachev's leadership was full of paradoxes (Hosking, 1990; White, 1991). He allowed some degree of decentralization by offering more autonomy to individual enterprises while, at the same time, setting up new institutions of centralization and discipline. He proposed that democracy was inevitable in Soviet society but could not give up the idea of a one-party regime. He criticized Stalin's methods but did not touch the problems of Leninism. And, as Andrey Sakharov (1988) pointed out, Gorbachev's reforms were, in fact, "a campaign to achieve democratic change by undemocratic means". Lewin (1988), however, argues that the contradiction of democratization in a one-party system was a problem only for western observers. For most Soviet people "democratization" did not mean the demand for a "western-style" multi-party system but rather an increase in citizen participation in political life. Thus the aim of Gorbachev's *perestroika* exercise was to get rid of the over-bureaucratic administrative character of the CPSU and return to the "original" party style which, Lewin (1988) argued, according to Gorbachev characterised the Bolshevik period before Stalin transformed it. Others, however, (Medvedev, 1980) argue that Gorbachev was neither a liberal nor a bold reformer. He only supported certain modifications and economic adjustments because his aim was to modernize and repair the system. At the same, he always rejected substantial structural changes, including giving up strict centralization, coercion or conservative dogmatism. The privileged elite remained unshaken in their unearned positions. Dogmatism remained dominant in intellectual life, and even economic and social progress would not continue unless radical changes followed them. Without the freedom of expression and a free press, the policy of "open borders" and open communication with the west, the Soviet Union could not achieve any further improvement. Thus Gorbachev's new type of "cult of personality" was typical of the Soviet-style of political system, argued Medvedev (1980). It was limited and initiated from the top down. If Gorbachev wanted "real democracy" in the Soviet Union he would have had to learn to share power. Gorbachev had a better intellect, better education and was a more decent person than many of his predecessors but he inherited a combination of orthodoxy and rigidity of the regime which made it more difficult for him to use political and diplomatic skills to manoeuvre in the Soviet political system. The one thing he did not even wish to pursue was sharing power.

The period was completed by the coup of August 1991 which removed one leader from power and replaced him by another one. Gorbachev (1997), in his account of the events of 1991, argued that the country had slid into a systematic crisis by the time of the coup. There was instability and chaos as a result of a mass of contradictions, and the reforms could not be carried out properly in a society which was used to decades of totalitarianism and state monopoly. Meanwhile the reforms created enough turmoil to cause much pain and unsolved problems in peoples' lives. One major problem with these reforms was that they were half-hearted and inconsistent. The Communist Party monopoly was not abolished, and the structure of the party bureaucracy remained the same. This is what provided the coup-plotters with a possibility of attempting to return the country to old-fashioned totalitarianism. "If this coup d'etat had happened a year and a half or two years earlier it might, presumably, have succeeded, but by 1991 the whole society, including the army and most of the people working for the KGB, had changed. The coup did not have a chance" (Gorbachev, 1997, p.123).

The coup failed because the public was ready to defend the country's fledgling democratic institutions and was certainly not ready to return to the previous regime (Smith, 1992). The coup played an important role in the history of the USSR: it created a situation in which the entire Union could not remain united any longer and collapsed into independent countries. The Soviet "disunion" as the *Guardian* put it (1991) created the birth of fifteen independent sovereign nations, the largest of which remained Russia (148 million people, 83 per cent of whom are Russians).

In sum, the Gorbachev period was a typical example of the way Russian reforms have been carried out throughout its history: from the top down, failing to incorporate the grassroots initiatives urging Gorbachev to establish a multi-party democracy. Nevertheless, even if Gorbachev became a brake on progressive development towards the end of his leadership, he was most certainly a very important catalyst in the history of the Soviet Union.

The First of the New Russian Period: Boris Yeltsin

The last coup attempt (by Rutskoi and Khasbulatov) in Russia provided strong evidence for the argument that Russia is politically not mature in handling disagreements by democratic political means. The lack of experience in democratic ways in the Russian political culture, due to long-term historic reasons, as we have argued throughout in this chapter, does not even today allow Russians to have faith in evolutionary, parliamentary means of development. Those in power are not trusted and, instead of replacing them by parliamentary means, armed "solutions" such as coups are sought. Although there have recently been fairly frequent elections in Russia (December 1993 and December 1995, legislative; and June 1996, presidential), political parties are weak and the system of political institutions is not developed. The consequences of the latest elections, the lack of development of democratic institutions at present, will be examined in detail in chapter 8.

Conclusion

In this chapter I have demonstrated that the lack of democracy in Russia is linked to a tradition of authoritarian rule. The origins of this tradition lie in several aspects of Russian history: firstly, the autocratic overcentralized style of rule, secondly, the bureaucratized administration, thirdly, the collectivist style of Byzantine Christianity and the strong link between state and church; and finally, the late serfdom and the consequent delayed economic and educational development accompanied by a weak middle class.

Even though the dominant tradition is linked to the lack of democratic institutions in Russia, we should point out the existence of a very limited but important democratic counter-tradition, initiated mainly by the so-called intelligentsia, which has always been restricted in its social base and marginal in its political influence.

One legacy of the Russian tradition is that all attempts at reform (from Peter the Great, through Stolypin, Lenin and Khrushchev to Gorbachev) have been initiated from the top down. A second legacy is that there is a belief among the Russian people in the need for a strong stabilizer in society and a tolerance of a strong state. A third legacy is that "democratization" in Russia means citizen participation in debates about the limits of power and bureaucratic nature of the state, rather than a strong commitment to party politics, which is weakly developed even today. These arguments will be demonstrated in the second half of the book.

The implication of my argument in this chapter is that the influence of the Communist period was due to a combination of the strong effect of the pre-revolutionary past, combined with the Marxist-based Bolshevik ideology of the dictatorship of the proletariat. In other words, I argue that the system under Lenin, Stalin and Brezhnev with its Communist ideology was built upon a pre-existing Russian tradition. Having reviewed the way democracy have not developed historically in Hungary and Russia, we now turn to the debate about whether opposition existed under Communism.

Notes

1. Christianity in Russia was established in AD 988. The previously united Christian Church split in AD 1054 and the Constantinople wing became the so-called eastern Orthodox Church as opposed to the Rome-centred Latin or Roman Catholic Church, which became the dominant denomination in western Europe before the Reformation. This division was the first to divide western Europe from the Balkans, the Christian part of the Middle East and Russia (Vernadsky, 1972).

2. At this time the United States had eight million, France twenty-six million and Britain eighteen million inhabitants (Seton-Watson, 1967).

3. Twenty-two million serfs, who had lived in bondage to 106,000 landowners, were liberated in 1861.

4. The three main advocates of these views were Bakunin, Herzen and Chernyshevsky.

5. The Socialist Revolutionaries was the party of moderate socialists, a mainly peasant-oriented party.

6. The Social Revolutionaries had 410 seats (out of a total of 707) of which 370 belonged to the right wing of the party. The Bolsheviks only had 175 seats (Mawdsley, 1987).

7. The Land Decree of 1917-18, for example, aimed only at large latifundia. The big estates were carved up resulting in a large increase of smallholdings. This meant the abolition of the last vestiges of a semi-feudal system. Although workers' control was officially extended to all industrial, commercial and agricultural enterprises (employing at least five workers and having a turnover in excess of ten thousand roubles per year) ownership was left in the hands of the original proprietors. Politically the term socialism was used prominently in the new constitution, adopted in July 1918, but it was intended to give proof of the determination to effect the transition towards socialism (Liebman, 1970).

8. The Ukrainians (about 32 million), the Belorussians (5 million) and the Poles (8 million).

9. Minorities, such as the Baltic nations: the Finns (3 million), Estonians (1 million), Latvians (half a million), and the Lithuanians (1.5 million); and others, like the Germans (1 million), Tatars (2 million), Buriats and Iakuts (together half a million), Armenians (1.5 million), Jews (3 million), Rumanians (1 million), Muslim Kazaks (4.5 million), Uzbeks (4 million), Turkmens (1 million), Georgians (2 million) and Tadzhiks (1 million), to name only the larger minorities, and further eighty other minority groups, each of which accounted for less than a million inhabitants, totalling at least ninety-six different nationalities, constituted an enormous problem of disunity in 1917 (Mawdsley, 1987).

10. Peasants, for example, were to turn over about a quarter of their crops to the government but could sell the rest in any manner they saw fit. The roadblocks, army and police checkpoints were abolished, and peasants were allowed to travel to towns to sell their eggs, meat, grain, and produce. Small manufacturers and larger scale enterprises reappeared producing consumer goods (Lincoln, 1989).

11. Steel production increased fourfold, coal production fivefold, and generation of electric power ninefold.

12. Malenkov, Bulganin and Nikita Sergeyevich Khrushchev emerged as new leaders (Nove, 1989).

13. At the famous twentieth Party Congress in his "secret" speech, February 1956. The speech became known as "secret" because it was not published immediately. Ironically it became one of the best known speeches ever delivered at a Communist Party Conference (Thompson, 1990).

4

Pluralism, Opposition and Civil Society in Soviet-Type Regimes: in Conceptual Debates and in Practice

In this chapter I will review the theoretical debates about the concepts of pluralism, opposition and civil society in societies with Soviet-type regimes, and the practice regarding opposition in the Soviet Union and Hungary, in the one-party period. It will be shown that, although both societies were categorised as "Communist" prior to 1989, the nature of opposition was very different in them. Let us start by reviewing the conceptual debates.

Totalitarianism and Pluralism in Soviet-Type Regimes: Concepts and Debates

The debate on pluralism in the Soviet and eastern European context was a response to the "totalitarianism literature" which was found one-sided and rigid by the advocates of Soviet-type pluralism. The totalitarian approach had a narrow, monistic view of Communist regimes. Based on the formal structure of the political system, it argued that, in a regime with only one political party and no free elections, there was no scope to express *any* opinion other than the prevailing one dictated by the Communist Party. Although the term originally referred to Nazi Germany and Fascist Italy, after World War II the Soviet Union was also identified as totalitarian. A large amount of literature concentrated on discussing totalitarianism applying it to the Soviet-type regime (e.g. Huntington and Brzezinski, 1964; Friedrich, 1966; Levi, 1966; Kassof, 1968; Meissner, 1969; Brzezinski 1969; Conquest, 1969;Brown, 1984).

Totalitarian rule was defined as "a form of personalised rule by a leader and an elite who seek to dominate both society and the regular, legal structure which is called the state" (Schapiro, 1972, p.102). Others (Friedrich and Brzezinski, 1966) argued that totalitarian regimes "remould" and transform their citizens in their image and ideology. The essence of totalitarianism is the regime's total control over the everyday life of their citizens, of their thoughts and attitudes. Totalitarian dictatorships, such as Communist states, consist of an ideology, a single-party rule (typically led by one person), a terrorist police, a communications monopoly, and a centrally directed economy. The apparatchiks are part of an extremely centralized and rigidly hierarchical bureaucratic organization with a high level of institutionalisation. The ruling party maintains unquestioned supremacy over the society, imposing its ideology at will (Brzezinski, 1969). The Communist system, according to Huntington and Brzezinski (1964), combined its high institutionalization with high pseudo-participation by individuals, a system in which people, including the young, became dull conformists. The party monopolized the function of integration by terror in Stalin's time and by means of bureaucratic arbitration after then.

Meissner (1969) emphasised three main characteristics of the autocratic-totalitarian regime. The first was unrestricted autocracy of the party, and the second was total control from above of all social organizations and institutions and also of all mass media and other sources of public information. Even when the period of all-encompassing terror was over, the control of all functions and thought in every section of society remained totalitarian (Meissner, 1969). The third characteristic was total planning, extended not only to the economic but also to the political and cultural sectors of society. The claim that totalitarianism did not cease to exist in the Soviet Union with the death of Stalin was echoed by Levi (1966) who argued that power within the party became concentrated in the Presidium after Stalin's death, and even more in the hands of Khrushchev, who exercised absolute control over all information and propaganda. Khrushchev regularly overruled even party bureaucrats to maintain his own leading position.

Even those followers of the totalitarian approach who recognised that after Stalin there was a certain degree of power-sharing among the leadership insisted on describing it in a demeaning way—from a "cold war" perspective—in accordance with the anti-Communist rhetoric: "The only significant reality in Russia is the "Byzantine" structure of top-level politics... As in the Jacobean tradition, a third-rate, faceless, collective leadership holds power in an increasingly nondescript society when their betters had driven each other out" (Conquest, 1969, p.66). The highly organized, strong and experienced bureaucracy, which was built by Stalin but continued to exist in the post-Stalinist Soviet regime, was organizationally highly effective at containing the social, political and economic forces, but was not designed to cope flexibly (Conquest, 1969). A few hundred people wielded as much social "weight", due to this totalitarian structure, as is usual for whole social classes, and operated with methods comparable only to those used by past despotisms (p.67).

The political changes in the Soviet Union after the Stalinist period, however, opened a wide-ranging debate over the question of whether the "totalitarian model" should be changed in the light of the changing reality in the Soviet Union and eastern Europe or whether it should be kept and attempts made to ascertain how far Soviet-type societies deviated from the "model". Some (Brown, 1984), although rejecting the totalitarian model, did so not because of the built-in political bias in the concept, but because the totalitarian interpretation of the post-Stalinist period had several shortcomings. Brown (1984) argues firstly, that the model exaggerates the success of political socialisation in Soviet society and implies that (a) the CPSU has been monolithically united and (b) that the party leadership managed to control all popular beliefs and values, which was not the case. Secondly, the model does not recognise policy processes and other political changes in the Soviet Union and other Communist states which were initiated from below.

Pluralists, however, emphasised a different point. They argued that *within* the framework of the one-party system there was a range of different views and interests which were expressed in the Communist-led regimes (Skilling, 1966; Hough, 1977). It was recognised that pluralism in the Soviet regime was different from pluralism in a multi-party system, but if pluralism was defined as: "interest-representation and group struggle by different sections of society" then it was also relevant in the Soviet regime (Hough, 1983, p.168).

The concept of pluralism, which was "rediscovered" by Skilling and others in the 1960s to apply to the eastern European and Soviet case, was originally used by British and American philosophers of the late nineteenth and early twentieth century (William James, John Dewey, G. E. Moore and Bertrand Russell) (Solomon, 1983). The British philosophers were concerned with the growing role of the state in Britain and argued firstly, that the conservative view of the state as legal, moral and political sovereign was false, secondly, that non-political groups and associations are prior to the state and hence had legitimate claims upon it, and thirdly, that the concentration of power in the absolutist state is an impediment to liberty. Liberty—they argued—is best preserved through the dispersion of power to groups and associations (Solomon, 1983). In the American literature pluralism became a frequently discussed term after World War II by writers such as David Truman, Robert Dahl, Nelson Polsby and Raymond Wolfinger. While the British scholars' main anxiety was rooted in an internal concern with the British state, which was steadily increasing its functions and its power, the source of American pluralists' interest was an external matter, the historic shock of two types of totalitarian regime, Hitlerism and Stalinism. Thus American pluralism in the 1950s concentrated on valuing and preserving the American status quo against totalitarianism. American pluralists shared the British view that the nation had to be fundamentally associational in character and groups had an important role to play in the political process, but they felt that the fragmentation of government produced a need for compromise with an emphasis on political participation (Solomon, 1983).

The study of pluralism in Soviet-type societies began in the post-Stalinist period with the early works of Skilling (1966), Hough (1977) and Hammer (1974). The new developments under Khrushchev made some "Sovietologists" recognise the existence of pluralistic features developing in the Soviet Union and the satellite states. It was argued (Hough, 1977) that there were major similarities between the American and Soviet-type of system regarding their "institutional pluralism", with the major difference that the American model of pluralism allowed the formation of independent pressure groups or parties while in the Soviet system those who wished to effect change had to work within existing institutions.

After the Khrushchev "thaw", however, when Brezhnev returned to a more rigid system in the Soviet Union, it was acknowledged that "bureaucratic domination" had again become strong (Hough, 1983). Hammer (1974) coined the term "bureaucratic pluralism" and argued that pluralism even in the United States is restricted by the existing bureaucracy and so this was not unique to the Soviet system, while Skilling (1966) noted that the Soviet Union had a "polyarchical system" meaning that it was "oligarchical rather than democratic in character".

The use of the term pluralism in relation to Soviet-type societies was often contested. It was suggested that it could only be used with qualifiers, like "limited pluralism" or "quasi pluralism". Others (Hough, 1983; Solomon, 1983; Brown, 1984) considered that the concept of *corporatism* is the most useful one in understanding eastern European societies. While the pluralist model emphasised conflicts between groups within society, the corporatist model talked about consensus and cooperation via the state (Schmitter, 1982). Contesting interest groups were recognised in the Soviet regime by Skilling (1971) who identified three politically active groups with different views. The first group contained officials and bureaucrats, apparatchiks and managers, the police and the military. The second group, like lawyers and economists, had some critical or independent views, but were not antagonistic to the regime and believed in helping by frequent consultations with the first group. Finally, the third group included people with independent or critical opinions, such as liberal writers, or opinion groups who were outspoken but often severely condemned. Skilling's (1971) definition of political groups was therefore wide, embracing a large spectrum of society, including Communist politicians, and concluded that "political groups in the Soviet Union are seldom organized, and if organized, are dominated by functionaries who are usually not elected and not responsive to the wishes of their constituents' (p. 382).

In sum, the most important element of the "pluralistic approach" was its treatment of Soviet-type societies as more complex than the "totalitarian" approach, which characterised Communist regimes in a very narrow way. In comparison with the monistic totalitarian approach the concept of pluralism within the Soviet regime was more useful because it distinguished between the different clusters of actors and interest groups and understood the process of decision-making within a Soviet-type regime in its diversity. The term pluralism, it could be argued, was not the most fortunate one because it could be misleading. Nevertheless, there was an important attempt to discuss the Soviet type of "multiplicity" and recognize that the system contains different social forces expressed by groups operating within and outside the Communist Party.

The Debate about Opposition in Soviet-Type Regimes

Recognising the complexity of different social forces within Soviet-type regimes, which were independent from the party organization, led to a focus on opposition in eastern Europe and the Soviet Union.

First, the argument concentrated on the term *opposition*. It was claimed that the term opposition could not be adopted in Communist regimes in the absence of rival political parties and, instead, the term "dissent" was suggested (Schapiro, 1972). Dissent did imply criticism and disagreement with the policy of the government but without any intention of violently overthrowing it. It has never been claimed, even by the "pluralist school", that the Soviet system was democratic in any way, but only that there were strong elements of opposition within the one-party state (Dahl, 1971). The debate concerning the term "opposition" drew attention to arguments such as Schapiro's (1972) claim that it originates in Lenin's concept to designate critics whom he intended to silence. As these critics did not intend to replace Lenin's administration by one of their own, but only wanted to criticize certain faults of that administration, their action should be categorised as "dissent", rather than "opposition", according to Schapiro (1972). Others (Medvedev, 1980) defined a dissident as someone who disagrees with the ideological, political, economic or moral foundation of a society but does more than simply disagree and think differently. She or he openly proclaims his/her dissent and demonstrates it by putting himself or herself in constant political danger (Medvedev, 1980).

According to Dahl (1966), oppositional forces were different from each other in the different democratic (i.e. multi-party) societies themselves, and this made it even more difficult to define the term within a one-party system. This inspired Skilling (1972) to re-define opposition in non-democratic countries and argue that opposition "has normally been forced to assume a variety of non-legal or illegal forms and to express itself in other than a formal and institutional manner" (p.73). Similarly Ionescu (1967) distinguished between *opposition*, which refers to a conflict of interest and values and incompatibility of opinions, and *political opposition*, which is institutionalized, recognised and legitimate. Ionescu argued that in Communist states while political opposition, as an institution, was denied, opposition existed. Political conflict in a non-opposition state formed a continuum of sui-generis situations and phases which inevitably led to institutionalized political opposition even if the intermediary situations remained stationary for a longer period. Ionescu (1967) identified four main groups differentiated by their motives. Firstly there were those whose political grievances concentrated primarily on the demand for freedom of expression of opinion and of information. Secondly there were those who had social and professional grievances and claimed that the government discriminated against them and obstructed their professional or commercial activities. The third group felt that the state interfered with their religious activities. The fourth group had nationalistic aims, and felt that their country or region was ruled by a suzerain power and felt oppressed by a central administration as ethnic or regional groups.

Four other types of opposition were identified by Skilling (1972). The first was *integral opposition*, which was based on Dahl's (1966) *structural opposition* defined in his book *Political Oppositions in Western Democracies*. Integral opposition meant overt or covert disloyalty, including underground activities or even revolutionary conspiracy. The second was *factional opposition*, which involved rivalry within the party or government. This did not imply opposition to the Communist system but involved fundamental ideological rifts between politicians. The third was the *fundamental opposition* of interest groups outside the circle of politicians. These groups lobbied the opposition within the *factional group* and sought to establish alliances and influence them to achieve changes in policies. The fourth type was *specific opposition* which referred to opposition within the system, for example, from inside the party opposing specific policies (Skilling, 1972).

Schapiro (1972) found Skilling's categorisation of types of opposition inadequate and proposed instead a five-fold classification. The first form was the complete ("all-out") rejection of the regime with a desire to overthrow it. Examples of the revolt in 1956 in Hungary and in 1967-68 in Czechoslovakia were cited by him. The second type was the *power struggle* or *factional conflicts*, when political leaders tried to oust each other. This was a frequent occurrence in regimes where conditions of intrigue and secrecy were present. The third category was "*protest* against the arbitrary abuse of law, procedure and civil rights by the Soviet authorities; against policies like the invasion of Czechoslovakia, or the oppression of national minorities;

against interference with freedom of speech and writing" (p.6). The fourth category included *interest group* or *pressure group* activities. Schapiro distinguished between these two groups by arguing that the former brought pressure on the government in order to promote its own interest while the latter sought to promote a policy beyond the group's own particular interest and put pressure with an aim of a more general nature. The last category was *pragmatic dissent* which included experts such as reform-economists, scientists, technicians who won a degree of freedom from party control after Stalin's death.

Bugajski and Pollack (1989) found Schapiro's categorization inadequate because he ignored the fact that the groups often overlapped and changes occurred over time. Further, although open dissent was restricted to a small number of people, while passivity and apathy characterised the majority, on occasions economic and social crises reached the point where people overcame their fears and undercurrents of discontent surfaced. Despite the strong party control "numerous sources of discontent persist[ed] among aggrieved social groups within East European societies and mushroom[ed] into open conflict under pertinent conditions" (Bugajski and Pollack, 1989, p.37). This process started after Stalin's death and occurred in several eastern European societies, such as Poland, Czechoslovakia and Hungary. Intellectuals and students began the process of discussion by acting as spokespeople, arguing for political reforms and criticising official policies. This influenced other social strata and sometimes even penetrated the ruling parties. Analysing the period of 1980s Bugajski and Pollack (1989) differentiated between (a) the national democrats, whose main concern laid in winning independence from Soviet-Communism, (b) the religious oriented Christian Democrats, (c) the liberal democrats who were pressing for individual liberties, (d) the social democrats who stressed the role of the state in improving the welfare system, and (e) the socialists who preferred a mixed type of economy with a combination of state ownership and privatisation in certain sectors, emphasising the importance of workers' self-management, traces of which originated in the pre-war and immediate post-war period, it was argued. Bugajski and Pollack's (1989) classification, however, could only apply to the last period of the Communist regime.

I argue that the different categorizations of political opposition in Soviet-type regimes have one major problem in common: none of them use more than one dimension to categorise the different groups. In contrast I identify two different dimensions, one of which has two aspects, along which oppositional groups should be categorised. The first dimension (A) which has two aspects (A1 and A2) measures (a) the level of resentment against the regime (A1) from high to low, from active opposition via apathy and cynicism to factional conflicts among political leaders and (b) the distance from the ideas of the ruling Communist party (A2). The second dimension (B) refers to the type of demands such as "intellectual type" or "abstract" and "working class type" or "concrete", following Szeleny's argument described earlier.

The highest level of A1 was found among the opposition which involved active political action expressing discontent. This was a small hard core of active dissidents who produced samizdat literature and organized demonstrations or strikes. The next highest level was found among the people who read and circulated illegal samizdat literature and participated in demonstrations, underground meetings, and illegal strikes. The next level included the so-called reformers who did not step outside the limits of legality but were strong opponents of existing policies. The fourth level on the A1 dimension included the large group of open critics who expressed strong disagreement with many aspects of the regime. This opposition became a regular feature of conversations and was more characteristic in Hungary than in the Soviet Union. The last category is "apathy and cynicism" which was very widespread both in the Soviet Union (Yanitsky, 1993b) and in Hungary (Kulcsár and Dobossy, 1988). This was, however, the "mildest" expression of resentment against the regime. The majority of people belonged to the two latter categories, but they should be described as a para-opposition, using Schöpflin's expression (1979), meaning that although they did not support the regime, they were politically passive.

The second aspect of the first dimension (A2) distinguishes opposition according to its distance from the ideas of the ruling Communist Party. Those furthest away from the party's views were the members of the hard-core opposition. This was a small number of people, as mentioned above, who openly and bitterly rejected the one-party system and all party ideas, and expressed their views in writing, published by the strictly illegal samizdat publications. They often lectured at underground meetings, smuggled out writings to the West and sought any occasion to express their complete opposition to the regime. Members of this small group of people were mostly well known to the authorities, often harassed by the police, lost their jobs and were sometimes sent into exile abroad. At other times they were denied passports to travel abroad and in the Soviet case (but not in Hungary) were sent to labour camps, asylums or into internal exile. They became "professional dissidents".

The second group was larger. These were those people who read the samizdat materials (which was an illegal activity) and circulated them with the deliberate intention of reaching as many fellow-thinkers as possible and those who made up the audiences of the underground lectures and who were participants in illegal demonstrations and strikes. They did not suffer too much harassment by the secret police though they were also black-listed and their telephones were often tapped.

The third group of people was the part of the population which was full of discontent against the regime and the government. They did not read the samizdat or go to illegal meetings, and did not do anything actively against the ruling power, but openly expressed their opinion. Their political expression went only as far as complaining and blaming the regime for everything. As this group made up the overwhelming majority of the population this created the basis of the lack of legitimacy which characterised the socialist regimes and contributed to their unexpectedly rapid collapse. The fourth group contained those people whose careers benefitted from the regime and who had some sympathy towards it though they had some criticisms as well. This was a large minority group. These people firmly believed in the regime and its core ideology but disapproved of certain aspects and practices which they thought needed to be modified. In their opinion the errors were due to the personal misconduct of certain individuals. These people later became labelled as "reform Communists".

Finally, was the very small group of hardliners, the closest to the party-line, who had privileged positions and strong loyalty towards the regime. They were in opposition only in the sense of contesting each other for even higher positions, more privileges and power. An example is the opposition which toppled Khrushchev. This type of opposition was also recognised by both Skilling (1972) and Schapiro (1984). The opposition of these people was mainly an internal affair amongst those who were closest to the government but it gained much attention among "Kremlinologists".

The second dimension (B), as explained, distinguishes between the types of demand. The intellectual opposition had an abstract approach to political grievances. They were mainly concerned with the lack of freedom of expression, freedom of the press and publication and freedom to organize independent pressure groups, movements and political parties. The working class opposition, on the other hand, concentrated on price increases and living standard problems and some of them demanded more workers' participation in management decisions.

In sum, I have identified two dimensions of opposition one of which had two aspects (level of resentment and distance from ideas of the ruling Communist Party) and the other of which referred to the type of demands. This classification will be used in the second part of this chapter when I compare opposition in the two societies.

The Debate on Civil Society

Having considered the débates on pluralism and opposition in Soviet-type societies prior to 1989, I now turn to the question of civil society, a term which became frequently used in the literature although it was used both in the eastern European and the western literature, often in different ways. Kumar (1993) pointed out that "using this distinction, East Europeans were mainly concerned with the construction—more or less de novo—of 'political society', for their idea of civil society was fundamentally one of social groups capable of self-organization independently of the state" (p.168). The term civil society is not used in the Marxist sense to refer to the private, nonpolitical sphere of the citizens' life, but more in the Hegelian and Gramscian sense to refer to activities outside the realm of the party and the state as well as the realm of the private life of individuals. It is associated with the political activities of people as citizens and their participation in civil and professional voluntary organizations or pressure groups, which have effects on their political socialisation as well as acting as mediators between the state and the individual.

Civil society has been the object of extensive debates (Keane, 1988; Cohen and Arató, 1992; Kumar, 1993; Bryant, 1993; Kumar, 1994). It has a centuries old history and has been applied in many different types of society from eighteenth century North America and Europe to present day eastern Europe. As a result the concept has undoubtedly acquired a "catch all" character, as Kumar argued (1993, 1994), especially in the recent eastern European literature where it is not only used in a different way but became is often used without being clearly defined. Hence before making analytical use of the concept it is essential to define it. The definition used here (and before; see Pickvance (1997)) is that civil society is the realm of political protest and civil activities, which is extra-parliamentary and which does not seek to gain power but to limit it. Thus civil society, in my definition, refers to those political activities which lie outside institutionalized state activities and party political activity, although the connections between civil society and the latter two types of activities are important, as argued by Tarrow (1983, 1989), Kriesi (1991), Habermas (1992) and Rootes (1992). The role of political activities within the domain of civil society is to protest and oppose, to counterbalance and limit power. This can be achieved by means ranging from petitions, demonstrations, and campaigns to social movements. It is the task of the representatives of the authorities to organize and regulate society but politicians need control. What is distinctive about civil society is this role of limiting political power, unlike opposition political parties whose aim is to gain power.

The problem of civil society was first approached by authors of the Scottish Enlightment, who argued that modern society breeds political despotism, and that therefore the creation and strengthening of citizens' associations, such as courts of law, citizens' militias and civil society at large, consultation, opposition and civilised persuasion can only bring protection against it (Ferguson, 1767, published in 1966). The most important synthesis of the concept of civil society was developed by Hegel, who analyzed it as a system of needs, of isolated individuals confronting each other in terms of antagonistic interests dictated by market relations. For Hegel the realm of the state consisted of civil servants, the police (*the authority*) and the crown (*monarchy*), while civil society contained the classes of individuals (*Stande*), the corporations (*associations*) and the umbrella organization of these associations: the *estate assembly*, and *public opinion* (Hegel, 1821). Hegel, like Montesquieu earlier and de Tocqueville later, recognised the need for an intermediate level of power between individual and state. The individual is powerless as an atomised subject vis-a-vis the state bureaucracy. The fear of despotism, which motivated Ferguson and Paine, however, was not present in Hegel's approach (Keane, 1988; Arató and Cohen, 1992).

De Tocqueville followed Hegel but based his views on the American context (*De la Démocratie en Amérique*, written in 1835-40) and an examination of the French Revolution (*L'Ancien Régime et la Révolution*). He highlighted the danger of the gradual concentration of power in the hands of a centralized administrative state, and argued that, while the state was supposed to regulate the conflicting particular interests of the different groups of civil society, it was actually becoming a popularly elected despotic power due to the very concentration of its power. According to de Tocqueville, a pluralistic, self-organizing civil society is the foundation of a democratic society, it is the *indispensable condition* of democracy. State power without the social safeguards of independent civil associations is a licence for despotism.

Civil society was the sphere consisting of unpolitical individuals united only by mutual dependence through the division of labour for Marx. Political life was monopolized by the state, he argued, which signalled the loss of community and the denial of meaningful citizenship. The individual in a modern society is atomised and depoliticized: "A person's distinct activity and distinct situation in life were reduced to merely individual significance" (Marx, 1973, p.287). The member of a modern society, he argued, is both an individual and a bourgeois, a participating citizen in communal affairs and a subject of political regulation. The separation of state and society is the cause of political alienation and the formation of voluntary associations are the expressions of particular egoistic interests, determined by the market.

Not all followers of Marx rejected the concept of civil society, however. Gramsci based his concept on Hegel's ideas, recognising the importance of civil associations such as unions, cultural institutions, churches, clubs, neighbourhood associations and the plurality of political parties, and continuing the Hegelian concept of corporations. Unlike Hegel, however, he located both family and political culture within civil society (Gramsci, 1971). Gramsci was aware of the totalitarian nature of the Soviet regime under Stalin and argued that the centralized state was becoming the greatest block to the development of a free society and civil society. Bobbio, commenting on Gramsci's works, emphasises that "civil society in Gramsci does not belong to the structural sphere [to the base] but to the superstructural sphere ... a fundamental point, which has not been sufficiently stressed" (Bobbio, 1988, p. 82). Cohen and Arató (1992), however, suggest that Gramsci's concept "rendered the whole doctrine of base and superstructure irrelevant" (p.145), because both civil society and state express the same principle and logic—to integrate civil and political society in the state—and that Gramsci argued that this reduction expresses one of two different principles, hegemony and domination. Anderson (1976) and Boggs (1984) add to the debate by arguing that Gramsci's distinction between state, and in particular parliament, which legitimately encompasses both coercion and consent, *and* civil society, in which legitimate coercion is absent, is too rigid and simplified. As Arató and Cohen (1992) stress, Gramsci should have recognised Hegel's concept of mediating institutions between civil society and state. Instead he developed the concept of an independent civil society, independent from both the economy and the state. Gramsci viewed civil society as the outcome and object of class struggle. Consequently the ruling social group, whether or not it is the bourgeoisie, will be the hegemonic one in any particular society (Gramsci, 1971). Gramsci nevertheless developed Hegel's concept of corporations, modernizing it by emphasising the functions of social movements, cultural institutions, civil associations and unions—as long as the working class is in opposition. But he treated many of these associations as pure vehicles for reproducing bourgeois hegemony, which must therefore be destroyed and replaced by forms of association which create a counter-hegemony such as alternative forms of associations, like workers' clubs and the associations of "organic" intellectuals who do not support the bourgeoisie but the proletariat, and parties of the proletariat itself. He believed, following the orthodox Marxist tradition, in the replacement of capitalism with another form of society via revolution and remained a life long supporter of the Soviet system even if he saw some of its contradictions.

Habermas (1992), does not dispute the relevance of the term civil society but explains how and why it has developed through history and how it has been transformed in different societies. As civil society in its modern sense emerged when capitalism emerged it will be useful to compare the period of the emergence of capitalism originally and now in the post-socialist situation. In the early period of capitalism a new class, the bourgeoisie, was emerging, which could not be assimilated with the nobles and courts any longer. In this new stratum "the state authorities evoked a resonance leading the *publicum*, the abstract counterpart of public authority, into an awareness of itself, as the latter's opponent, that is, as the public of the now emerging *public sphere of civil society*. For the latter developed to the extent to which the public concern regarding the private sphere of civil society was no longer confined to the authorities but was considered by the subjects as one that was properly theirs' (Habermas, 1992, p.23). Thus civil society for Habermas consists of two realms: the public and the private spheres.

Habermas's concept of the "public sphere", however, unlike my definition of civil society, includes political parties and parliament. This is because originally parliament's function was to counterbalance the authority of princes and nobles. However, Habermas himself admits that "from the very start, indeed, the parliament was rent by the contradiction of being an institution opposing all political authority and yet established as an 'authority' itself" (p.233). In the contemporary context, on the other hand, the role of parliament and political parties is different in both western and eastern European societies. Habermas also emphasises that, on one hand, private and public spheres in modern societies are indissolubly connected and, on the other, that the private sphere is reduced to family life and leisure activities in contemporary societies. Consequently what we are left with is civil society as the extraparliamentary polity.

Keane (1988), a contemporary interpreter of the civil society concept and its relation with the state in the current western European context, argues that its popularity is due to three major factors: the restructuring of capitalist economies, following the post-war boom and subsequent recession which leads to a permanent "mismatch" between the economic and the political spheres; the political controversies following the failures of the Keynesian welfare-state; and the rise of new social movements. Keane (1988) stresses that there is a constant threat to western civil societies from the activities of the state and the private corporations. He uses an ideal-typical approach in his distinction between state and civil society aiming to explain socio-political realities by analyzing particular institutions (their origins, development and transformation), or whole social systems.

The term civil society was used by the political right as a synonym for "private" life and freedom of market activities as opposed to state intervention. The orthodox marxist left rejected the idea of civil society on the basis that it does not address "fundamental" problems of property, class and class conflict. Some identify it entirely with a Gramscian approach and accept that it could only be used in his original understanding. Others analyze civil society from the point of view of social movement formation (Melucci, 1989), or the relationship between labour market, welfare state and the household (Offe, 1984), or use historical analysis (Elias, 1988; Szücs, 1988; Vajda, 1988; Habermas, 1992). Neo-marxist and neo-Weberian perspectives attempt to incorporate both state and economy centred approaches when analyzing civil society (Skocpol, 1979; Jessop, 1990; Offe, 1984; Giddens, 1985).

Arató (1991), who analyzes the question of civil society in eastern Europe, defines it as a sphere of social interaction between economy and state, composed of associations and publics. He considers it important to link the concept of social movements to that of civil society and argues that independent collective actions, citizen initiatives and social movements are all present in the transforming new societies, and should be distinguished from "political society" which refers to political organizations, parties and parliaments. Modern civil society is created through forms of self-constitution and self-mobilization which are institutionalized through laws. But civil society is not all of social life outside the administrative state and economic processes. This is because these are directly involved with state power which they seek to control and/or obtain. Civil society, on the other hand, is not directly related to the control or conquest of power. It aims to exert influence through democratic associations and unconstrained discussions in the public sphere. However, "such a political role is inevitably diffuse and inefficient. Thus the mediating role of political society between civil society and the state is indispensable, but so is the rooting of political society in civil society" (p.198). Concerning the eastern European and Soviet scene, Arató (1991) emphasises that a certain degree of economic development, growing consumption and to a certain extent tolerated depoliticization of the private sphere went ahead in the Soviet Union and most eastern European societies. He refers to arguments, such as Lewin's (1988), who stresses that modernization was responsible for the expansion of civil society in the eastern bloc. In Arató's view Lewin overestimates the role of modernization in the development of civil society and argues that the existence of social processes and relations independent of the state is a necessary but not a sufficient condition of modern civil society. He argues that Stalinism destroyed civil society and prevented it from developing, but that modernization contributed to its development. However, the pattern of modernization in the Soviet Union was in many respects a failed and even pathological one which endangered for some time to come the building of a genuinely modern political culture. If modernization were the only precondition of the emergence of modern political culture or civil society, the level of democratization and outlook for it would be much better than the present situation suggests in many countries around the globe (Arató 1991).

So far we have reviewed the arguments about pluralism, opposition and civil society, and their relevance in Soviet-type regimes. It was shown firstly that, even though the term pluralism was controversial, the pluralist approach to Soviet-type regimes was more useful than the monolithic, totalitarian one. Secondly, I have explained that many writers agree that opposition existed in the one-party system but debate the way to describe the different groups involved in opposition. Thirdly, it was shown that civil society is a concept which has been used both in the west and in the eastern European context, where it was especially useful during the period of transition. Its relevance has been debated both within western liberal democracies and in eastern European societies. I will return to this subject at the end of this chapter. Now I turn to pluralism and opposition in practice in Soviet-type regimes.

Opposition in the Soviet Union

First I will discuss the general political atmosphere towards the opposition in the Soviet Union under different leaderships.

Under Stalin oppression was tight, and the country isolated and strictly controlled. Large numbers of people experienced extremely harsh treatment from the regime and became victims of it even without participating in oppositional activities. After the death of Stalin, during the Khrushchev period (1953-64), a number of people joined vigorous debates on public policy matters. Experts and specialists, especially the cultural, professional and scientific intelligentsia, were invited to participate in decision-making. The intelligentsia thus gained a say in policy matters. Under Brezhnev the regime became more bureaucratic and politically less tolerant. The groups in opposition, as well as former advisers, who emerged under Khrushchev, became isolated again. Many of them faced arrest and subsequent trials where they were charged with "subversive" activities or anti-Soviet propaganda (Simmons, 1955). But the process, which started off under Khrushchev, turned out to be irreversible. Numerous groups were in existence and noted in the literature. Bilocerkowycz (1988) reports about the Ukrainian dissidents, often writers, but political groups as well. One of them was the Ukrainian Helsinki Group, an open and public organization, formed in November 1976, to defend national and human rights. Several other human rights organizations were established in the USSR, one in Moscow formed in May 1976, and others in Lithuania, Armenia, Georgia and Latvia. Official harassment of these groups began the very day they were founded (Bilocerkowycz, 1988). Friedgut (1979) provided important evidence of previously little known community self-help groups. These self-help organizations fulfilled the requirements defined by western scholars, such as Milbrath and Goel (1965), Verba, Nie and Kim (1971), and Kornhauser (1960). They facilitated the articulation of interests and served to recruit people into politics. Friedgut found examples of numerous community self-help groups which were not only tolerated but even supported by the local authorities. This was a widespread and well established phenomenon in the Soviet Union (Friedgut, 1979). The importance of these self-support groups of the Brezhnev era was that during the *perestroika* period these "tolerated" groups often became "hotbeds" of opposition activity, the so-called *neformaly* (informal groups) (Berezhovski, 1990). They turned out to posses important potential for future opposition when the political atmosphere allowed their existence as "proper" self-interest groups. Under the Brezhnev period they could accumulate "expertise" in organizational skills and (limited) self-support which was very necessary in order to engage in mobilization when the political environment changed under Gorbachev.

For researchers of the Soviet Union the problem of the 1960s was to uncover the complexity of Soviet society and distinguish between several interest groups to reveal its pluralist nature. By the 1970s, however, it became clear that there were quite a number of people in the Soviet Union who openly expressed considerable unease or even discontent with the regime. This became known in the literature as the "Soviet-type of opposition". There were several groups within Soviet society which were clearly distancing themselves from the party propaganda. Firstly, there was a group of people who were better educated, and who included social scientists, writers, artists, social workers, scientists, etc. Secondly, there was the younger generation which felt that the Brezhnevian policy of isolation cut them off from the "west", including pop culture, and rebelled against it. Clubs, like the "Independent Song Club" closed by the authorities in 1975, or the "Student Club" organizing double meaning pantomimes, again later banned by the authorities (Mandel, 1989), were examples of activities of younger people, often students, who were feeling deprived of access to western culture and created their own culture, which was not in line with the official socialist direction in popular culture.

The Gorbachev era (1985-91) turned out to be the last Soviet period and brought an unprecedented degree of political tolerance to the Soviet Union. Whereas before this the regular conflict between government and opposition had led to direct confrontation, under Gorbachev a more tolerant attitude developed. Civil society, dormant or "disguised" under Brezhnev, sprung up and direct repression by the authorities became much less frequent. The process of de-totalitarianization and democratization had been started by reconstructing the constitution, restructuring political institutions (Sakwa, 1993), and the initiation of a western type of pluralism (Tismaneanu, 1990).

I will now examine the development of opposition among people concerned with ecology, workers' and women's problems, and peace prior to the regime change. Environmental movements will only be touched on here since the development of environmental ideas as well as movements will be discussed in chapter 6 for Russia, and in chapter 5 for Hungary.

The general political discontent in the Soviet Union was an important and widespread phenomenon but many people saw problems in more concrete terms and political activity sprang up around those more specialized subjects.

Ecological Groups

The ecological groups were mainly concerned with issues like the Siberian rivers Ob and Irtich, the drying up of the Aral Sea in Kazakhstan and protest against the nuclear pollution of the Caspian sea. The most influential, and perhaps best known, ecological movement was, however, the long standing protest against the pollution of Lake Baikal. This case will provide us with a good example of "Soviet-type" opposition operated in practice.

The large Siberian lake was chosen in the 1950s as the "best" site for industrial "progress" (Komarov, 1980), to exploit the assets represented by the water, the surrounding forest, and the available local labour force which by background was mainly a poorly educated minority group of Buryats. By the 1960s this had resulted in beginning of the construction of two huge paper and pulp combines, the Baikalsk and the Selenginsk.

Although the history of the public protest to save the lake started in 1963, evidence suggests that, in fact, many experts, including biologists, hydrologists, and geographers had opposed the construction of the two combines from as early as 1958 (Komarov, 1980). The scientists' opposition, however, turned out to be too weak and ineffective compared with such powerful political forces as Gosplan (the State Planning Committee of the USSR), the "Committee on Forestry and the Paper and Pulp Industry" and the Ministry of Defence, and the construction was successfully completed.

In order to calm the first voices of alarm the authorities denied any access to the project-documents and the so-called "scientific reports" produced by the authorities, some of which went as far as arguing that the sewage water would create conditions for the propagation of life which would even increase the fish reserves.

Although public pressure grew both among local people and scientists, resulting in the 1960s protest letters signed by a string of prominent academicians and several well known Russian writers, the official Soviet press was never allowed to give a platform to the opposition, and the scientific lobby seemed to be ignored.

"Soviet-type" opposition, however, resulted in a number of steps being taken by officials. As early as 1966 it was announced that an especially powerful and sophisticated purification system had been installed in the paper combines and that a special "commission" had been set up to monitor the environmental effects. The official report, of course, found the filters to be "on the whole adequate". In addition, an official propaganda campaign was initiated, with films produced and articles placed in the official press to reassure the public.

The official propaganda policy, however, could only be effective for a limited period of time. By 1977 several reports were produced by academicians of the Academy of Sciences of the USSR. Clashes between the Soviet authorities and the members of the scientific communities, supported by many of the local population, became increasingly visible and well known.

The new type of "people's power", in the form of organized protest groups, accelerated in the perestroika period resulting in a number of movements. At the peak of this, in 1990, there were at least five known opposition groups in the region simultaneously campaigning for the lake. These included the Baikal Eco-World, the Baikal National Front, the Society for the Defence of the Baikal, the Centre for the Ecological Defence of the Baikal Region and the Baikal Fund (Stewart, 1992; Wilson, 1993). There were numerous demonstrations in the streets of Irkutsk and at the site of the controversial construction of a pipeline which was planned to divert the effluent from the Baikalsk paper mill into the nearby river Irkut (about forty miles away) instead of the lake, thus redirecting pollution to another site which has its own population and is also a popular holiday area. As a result the activists successfully organized a petition-campaign (with around 107,000 signatures) and there were several demonstrations. These activities were effective in two ways: firstly, the pipeline plan was abandoned and secondly, their efforts unified the previously scattered protest groups into a well organized environmental movement, the Baikal Fund (Wilson, 1993). The Baikal Fund, like all the other protest groups around Lake Baikal, is a well known environmental movement which came into existence because of the lack of concern over environmental protection by the industrial growth-oriented Soviet leadership. The Baikal Fund's success and popularity, however, is not only due to the "newly emerging Russian green movement" or "western scrutiny and international recognition", as Wilson (1993, p.65) argues. It has a lot to do with the Russian nationalist current of opinion, attracting, on the one hand, a number of well known public figures (like the writers Valentin Rasputin and Sholohov, or the economist Gennadi Filshin), and also mobilising romantic-nationalistic sentiments which could easily lead to a high profile national focus on the plight of an important national lake, with a strong emphasis on the preservation of "mother-nature". As Dunlop (1983) argued, seeking to safeguard Russian historical monuments and the environment from destruction are tendencies which stem from one root. They both exhibit Russian conservative "patriotic" thinking among both Slavophiles and National Bolsheviks, the philosophies of *vozrozhdentsy* and the Russian *narod*.

Evidence also suggests that there is a strong "nationalistic bias" when weighing the importance of environmental danger within the same region. Stewart (1992) pointed out that, beyond the Urals air pollution is all too evident "with chimneys emitting grey or black smoke producing up to several kilometres long smoke bands, causing serious health problems among children in Ulan-Ude"[1] (p.227). Yet there seems to be much less concern about the Buryat childrens' respiratory problems than about the state of this "very Russian" lake, the Baikal.

Concern over Lake Baikal thus became probably the most famous environmental protest movement in the Soviet Union, attracting great attention and mobilising huge opposition. However, this was not the only "green" protest which existed in the Soviet era. There were protests in the Moscow suburb of Kuskovo from 1979 concerning dangerous waste from a nearby chemical factory, which was finally closed in 1987 as a result of this movement. After the Chernobyl catastrophe there were several protest activities in the Ukraine and elsewhere in the country and a general questioning of the whole Soviet nuclear energy programme (Hosking, 1990).

Ecological groups often had in their leadership well-known writers or scientists whose fame and reputation helped to create the necessary publicity and raise public awareness. Ziegler (1991) argued that there were three important underlying elements in the rapidly emerging green movements in the 1980s in the Soviet Union. Firstly, environmental issues mobilized previously politically inactive citizens and they became focal points of voluntary political participation. Secondly, their important but implicit anticommunist message questioned the "growth at any cost" type of Soviet attitude towards the economy, resources and the environment. Thirdly, green movements were against centralization, rejecting the continuation of Moscow-centred policies and giving support to the idea of regional control over resources and environmental consequences. This notion often coincided with ethnic, separatist and nationalistic feelings (Dunlop, 1983).

Women's Groups

Feminism in the Soviet Union was not very widespread. But there were politically active women, who drew attention to the problems between the sexes and discrimination against women. Valentina Tereshkova (1987), the first woman cosmonaut, was active in the Union of Soviet Women and lectured about several unsolved problems Soviet women had to face. She emphasised that many women were still employed in heavy manual jobs in the Soviet Union, that women did the vast majority of queuing and other household tasks, that women occupied lower prestige jobs and were lower paid. She also drew attention to the high rate of abortion among Soviet women due to the lack of other contraceptive methods and the lack of family planning, and to the appalling and life threatening conditions for abortion in the hospitals which Soviet women had to resort to. Organizations like *Women and Russia*, which operated in Leningrad from 1979, or the *Union of*

Soviet Women united politically active Soviet women. From 1988 onwards a few, mainly professional, women formed groups such as the International Press Club of Women Journalists in Moscow, a women film-makers' union, a women writers' club, an association of women engineers and women scientists and most importantly the League for Society's Liberation from Stereotypes (LOTOS) challenging the traditional sex-role ideology. These women's groups, however, remained fairly isolated and the lack of gender consciousness posed a considerable barrier to the development of the women's movement in the Soviet Union (Nechemias, 1991).

Workers' Groups

Workers often protested against the conditions they had to work in, even under the harshest period of Brezhnev's rule. The strikes in the hydro-electric plant in Vyshchorod near Kiev, in the rubber plant in Sverdlovsk, in the armaments factory in Gorky in 1969 and in Vladimir in 1970, in factories in Dnepropetrovsk, Kopishche, Vitebsk, Togliattigrad in the 1970s, and in a motorcycle factory in Kiev in 1981 are a few examples of the frequent and growing expression of serious discontent among workers concerning their working conditions, the hours of unpaid overtime work due to poor organization and mismanagement, the poor safety and hygiene conditions, the high level of work accidents, and the lack of improvement in living standards. From 1975 the growing use of moonlighting and black market activity were only some of the spontaneous reactions by workers to their dissatisfaction with the regime (Hosking, 1990). Sedaitis (1991) argues that workers were courted by both the conservative and the reform-oriented politicians during the 1980s, each claiming to represent the Soviet working class. Workers themselves, however, often turned away from the larger political issues and concentrated on more concrete and immediate labour problems. They established unofficial unions and clubs in their factories or group of factories to advance their rights, raise their wages, demand more involvement in the factory management and call for more decision making power for the workers' collectives. Some demanded the depolitization of workplaces and the abolition of any prerequisites not related to job performances (Sedaitis, 1991).

Peace Movement

The Moscow Group to Establish Trust Between East and West was the most important unofficial peace movement in the Soviet Union. It was created in 1980 and called for genuine detente. Although the "Trust" group tried to avoid open criticism of the Soviet military build-up it was not seen as impartial at the time. Participation in the movement was considered severely subversive and the members of the Trust were arrested, deported or locked up in psychiatric hospitals. But the group from the Brezhnev era survived and by the Gorbachev period groups with similar platforms were mushrooming in the Soviet Union (Kuznetsov, 1990; Tismaneanu, 1990).

The Pattern of Soviet Opposition

Having looked at the opposition which developed in a number of fields, I now turn to the overall pattern of opposition. The sixty years of bureaucratic dictatorship created a general awareness of social ills, which was of course fragmented into different social milieus, but reflected a general trend. The deepening contradiction in the Soviet Union led to a gradual appearance of a consciousness. "A real public opinion has taken shape in this country" (Mandel, 1989, p.76) which led to the birth of civil society. This was in its embryonic stage in the 1960s but a considerable growth in civil activities occurred in the 1970s, accelerating in the mid 1980s (Hosking, 1990).

The economic and social stagnation of the Brezhnev years led to the degeneration of Soviet society into corruption and sloth while the isolated leadership promoted a cult of glory for itself, widening the gap between reality and triumphant empty claims of "advanced socialism". The result was massive cynicism and a popular rejection of the official ideology, especially amongst the intelligentsia and the youth. Surveys indicated that the government's extensive agitprop efforts did not result in the desired aims as the majority of the populace ignored the official propaganda (Smith, 1992). The rejection of the Brezhnevian style of regime should not mislead us, however. According to opinion polls (Smith, 1992) Soviet people developed a strong sense of egalitarianism, opposed large income differences and preferred a system in which the state guaranteed the quality of life by providing jobs, housing, health care, education, and pensions for its citizens.

The last period of the Soviet regime, the Gorbachev era, developed a number of reforms, even in the sphere of the economy, which encouraged the growth of independence from party control. The appearance of the so-called "cooperative" movement in 1988 (Slider, 1991) is one example of this. The Law on Cooperation, issued in June 1988, provided the legal framework for "cooperatives" to become the first small business enterprises, to operate without any direct Communist Party involvement in their activities. This was mainly because they were too small to have a local unit of the arty to control them. This made the cooperatives the first economic units relatively isolated from direct party control. They were obliged to look after their own interests, independent of state or party agencies, and were answerable only to their members. By 1989 the cooperatives joined together into national organizations in order to play a more direct political role and stand up for their own rights. They sought to influence economic policies and protect themselves on the national and regional level. Thus the so-called "cooperative" movement contributed, in the sphere of the economy, to the growing grassroots movement under Gorbachev.

A survey of young people conducted in 1987 showed that 65 per cent of young workers and 89 per cent of students of vocational and technical schools considered themselves to be members of illegal, so-called 'informal" or "independent" groups or _neformaly_, using the Russian word (Tolz, 1990). It is suggested (Igrunov,1989; Berezhovski, 1990; Yanitsky, 1993a) that about 60,000 _neformaly_ existed in the Soviet Union by 1989. When on 13 March 1990 Article 6 of the USSR constitution was abolished, the Communist Party lost its 70 year old monopoly on political power. This decision legalised a situation which had already developed by then. Different groups and factions emerged throughout the country demanding legal status. The various democratic and national organizations were in the process of organizing themselves into political parties, and gained strength and support by the day (Smith,1992). The first embryonic political parties in the transition from the Soviet period to the Russian era were born in 1989.

In sum, Soviet opposition came into existence with the Khrushchev "thaw" and although under Brezhnev the level of political tolerance was much lower, organized political opposition continued to exist. There were self-support neighbourhood groups strongly encouraged by the officials and a number of types of opposition group concerned with issues ranging from ecology and workers' rights to religious and nationalistic demands. One reason why these groups were not always widely known was that the media were strictly controlled under Brezhnev, and information about them was blocked. In Gorbachev's time this changed, contributing to the rapid growth of grassroots activities. Opposition in the Soviet period did not, however, necessarily mean being anti-socialist. Soviet people developed a strong sense of socialist values which they maintained even when opposing many features of the Brezhnevian period.

Opposition in Hungary

The case of Hungary was fundamentally different from that of the Soviet Union for several reasons. Firstly, Hungary had a much shorter period of Communist rule, secondly, it experienced the revolution of 1956 and thirdly, not independent from the events of 1956, the political and economic system in Hungary became different from those of most neighbouring socialist societies. In 1968 the New Economic Mechanism was introduced and from the mid 1960s up to the end of the socialist system political tolerance was greater and more consistent than in the Soviet Union. This resulted in much wider opportunities to express open opposition to the regime. It was possible to publish the kind of critical articles and books which in the Soviet case was impossible, due to the much greater control of communication there. This is why in Hungary the focus of opposition shifted towards publications instead of organized action.

The theoretical debates concerning the Soviet case mostly applied to the entire Soviet bloc. Meanwhile of course it was always acknowledged that eastern Europe was different from the Soviet Union. It was research on eastern Europe which prompted the term "pluralism" in Soviet-type societies in the first place (Skilling, 1966; Solomon, 1983a). This was the result of the recognition of the existing plurality of various interest groups in these societies. Later pluralist-literature incorporated the Soviet Union as well. But opposition was different there. Compared with the USSR, eastern European opposition demands were more centred on the concept of individual rights. This makes an interesting contrast with the west where demands for social rights increased after the war. In most liberal democracies after World War II the expectation was that the state should intervene using its administrative and judicial power to redistribute income and pursue certain social and economic rights for each citizen. Tőkés (1979) argued that this development in liberal democracies was influenced by the historic events in the less developed part of the continent and the world. The rise of social democracy, the trade union movement, and later the anti-colonial and national liberation movements were not independent of the leverage of the Russian Revolution and eastern European development after World War II. They provided strong precedents for the demand for expanded social rights in western liberal democracies. This, however, in turn influenced the more open central eastern-European societies, including Hungary, in their political demands (after the short spell of the Stalinist oppression) (Tőkés, 1979).

After the Communist takeover in 1948, a one-party hegemony was established in eastern Europe, led by hard-line Stalinist Communists who were trained in the Soviet Union under the great dictator. The Hungarian version of the "cult of personality" was headed by Matyas Rákosi, who himself spent decades in the Soviet Union prior to his return in 1945. Judging by the spectacular defeat of the Communists at the last free election in 1947 the majority of the population was against the Communist rule. Those groups of people who were going to lose their assets, position and influence were strongly against replacing a wide democratic coalition with Communist rule. They included the aristocrats, the bourgeoisie, those who were educated before the war and occupied managerial positions, and landowners. The churches were also strongly antagonistic to an atheist regime.

The lack of trust was mutual. It was, however, not restricted to these groups mentioned above. According to the slogan of the period—"the enemy is amongst us"—those who were not "trustworthy" gradually extended to include the majority of the population, just as in the Stalinist Soviet Union, which by the end included ordinary peasants and factory workers. In contrast to the Soviet Union, however, the shorter period under Communism created the situation where previous members of other political parties than Communists remained present in Hungary. They included Christian Democrats, Smallholders, former Social Democrats as well as "western" Communists, i.e. those who were not educated in Stalinist schools in Soviet exile but had lived either in western countries or in Hungary prior to 1945. All these people became targets of political harassment by the Stalinists. Between 1948 and 1956 political parties other than the Communist party remained "dormant" and their former members were subject to strict police and ÁVO control.[2]

The most important opposition in Hungary occurred in 1956. After Stalin's death in 1953, the Soviet backing of the hardliners weakened somewhat which resulted in division within the Communist leadership. Imre Nagy, a reform Communist, became prime minister in 1953 and demanded both political and economical changes. A bitter power struggle developed between the Stalinist Rákosi and the reformist Nagy between 1953 and 1955, with the majority of the public supporting reforms. When the first mass demonstrations of the 1956 uprising started on 23 October, public pressure persuaded the then prime minister, András Hegedüs, to resign and Nagy took over leadership of a new government which consisted of both non-Communists and a few Communists. The one-party system was abolished on 30 October 1956. Nagy demanded the Soviet troops' removal from Hungarian territories and decided to withdraw from the Warsaw Pact on 1 November 1956 (Grzybowski, 1991). The uprising ended on 4 November 1956 with heavy involvement of Soviet troops, and Nagy was subsequently hanged.

The new leader, János Kádár, was a pro-Soviet reform Communist, whose first aim was to re-establish the one-party system and close ties with the Soviet Union, but who combined this with economic reforms and political liberalization. As a result the critical opposition also appeared on the scene. This critical opposition was not, as Schöpflin (1979) put it, "a group of individuals, acting in a more or less organized fashion, who have mounted direct or indirect challenges of their governments by seeking to exert pressure on these governments for specific policy objectives" (p.142). The majority of the opposition during the Kádár regime was not trying to act outside the system and bring pressure on it from outside. Apart from two major pillars, namely Hungary's alliance with the Soviet Union and the leading role of the party, which were untouchable subjects, people were allowed to express relatively broad criticism within and outside the party as well. Public debates were frequent and encouraged. Publications, especially those with smaller circulations and away from central concerns, could carry articles airing highly sensitive points of view. This was the part of the critical current in Hungary which Schöpflin (1979) called para-opposition. The para-opposition "does not overtly question the ideological basis of the system, but does accept the leeway for a semi-autonomous political role permitted by that system" (p.142). In the mid-1960s three types of publication were defined: the supported, the tolerated and the prohibited. This became known as the policy of the three Ts from their Hungarian equivalent (*támogatott, türt, tiltott*). These categories were more guidelines than precise definitions and were subject to changing interpretations over time. Self-censorship by writers and editors therefore became a complicated and sophisticated system often based on intuition. This system, however, allowed a much wider range of political opinion to appear in Hungary than in most Soviet-type regimes and especially the Soviet Union itself.

Apart from this para-opposition there was a small but strong core of "real" opposition in Hungary. These were often individuals or groups of individuals rather than organized movements. The so-called New Left or as it is became known the "Budapest School", was the first to gain coverage in the western press (*Telos*, 1978). A strong public protest was triggered by the invasion of Czechoslovakia in 1968, involving Warsaw Pact troops including those of the Hungarian Army. György and Mária Márkus, Ágnes Heller, Vilmos Soós and Zádor Tordai put themselves in the vanguard of these protests by signing the so-called Korcula letter, but many others joined them, especially university students, and even within the Young Communist Organisation (KISZ) many campaigned in support of the Czechoslovak reforms and against the invasion. As a result of their open protest a group of seven people were denounced by the Central Committee of the Hungarian Workers' Party in 1973, including András Hegedüs (the former prime minister turned leading opposition figure), Mihály Vajda, János Kis, György Bence and those named above. They were dismissed from their jobs and for a long while were banned from publishing in Hungary. All of them were pressured to go into exile and many of them did. The next person to be persecuted was Miklós Haraszti for his book entitled *Piece-Rate in the "Red Star" Factory*, which was later published in English under the title *The Worker in a Workers' State* (published by Penguin in 1977). His trial aroused widespread support in Hungary. In 1973 there were also large scale petitions against the tightening of the abortion law. In 1974 two people were singled out on the basis of their political attitude based on their academic research and expressed in their publications: Iván Szelényi and George Konrád. Szelényi was forced to leave the country and Konrád left for several years (Schöpflin, 1979). In 1977 support for the Charter 77 movement was expressed in petitions.

Szelényi later argued that the majority of those in opposition were socialists. Their socialist values, however, did not coincide with those practised by existing state socialist societies. Being part of the socialist opposition meant being against totalitarianism, police oppression, censorship, restriction of civil rights, freedom of speech, and assembly but still having socialist views. The opposition were often highly educated intellectuals, but not exclusively so (Szelényi, 1979). Many people were dissatisfied with the regime and demanded reform socialism. A group of intellectuals demanded existing practices to be replaced by original Marxist principles. Others, however, were strongly against the one-party regime. The largest section of the population, however, opposed the regime on a non-intellectual basis. Their grievances were based on dissatisfaction with lower living standards than in developed western societies. Practical interests rather than abstract political considerations were the sources of grief for most.

There was significant resistance among workers against state socialism, which was expressed in a different way from the intellectual type of opposition (Szelényi, 1979). Intellectuals argued mainly through their published (or unpublished) writings and in underground meetings, discussions and legal or illegal lectures. Workers, on the other hand, used methods such as slowing down production, going on strike or simply cheating the regime wherever they could, for example by stealing public property. They were less articulate and their activities were often more spontaneous; workers might have lacked a coherent set of goals, but that is not to deny the strength of their discontent and should not lead us to underestimate the significance of their opposition to the regime. Commenting on working class opposition, Pravda (1979), Bugajski and Pollack (1989) also argue, that in general it was motivated by practical everyday concerns, material issues, like price increases, stagnating or decreasing living standards and rising expectations which could not be satisfied, or feelings of powerlessness and under-representation in the management of their enterprise.

Although the Communist leadership desperately tried to play down workers' oppositional activities, portraying them as a small minority of decent law-obeying citizens manipulated and misled by a few troublemakers, in fact, there were reports even in the legal literature of relatively widespread workers actions (Héthy and Makó, 1978). There were also demonstrations, which became widely known even though the media, being censored, did not report on them. One of these well known demonstrations took place in 1969 in "Red" Csepel, an industrial area on the outskirts of Budapest just after the introduction of the New Economic

Mechanism, which resulted in radically increased wage differentiation between workers and management, which workers were opposing. The unexpected political pressure made the political leadership reverse their decision and the new bonus system was abolished immediately.

People in eastern Europe developed a strong aspiration for steady material improvement, social mobility and effective social services, but these desires remained unfulfilled in most cases. As a result of the growing proportion of better educated skilled workers, an increasing demand for workers' autonomy and participation in decision making also developed. It often coincided with the prospect of underemployment and greater wage differentiation on the one hand, and more demanding work quotas without material compensation, stricter work disciplines or the prospect of unemployment, on the other (Haraszti, 1977).

The question then is why it was that intellectuals and workers were mostly separated and did not unite in opposition? One explanation was that intellectual dissidents become marginalized from the working class because the workers did not have appropriate organizations and therefore it was impossible to establish contact with them (Rakovski, quoted in Szelényi, 1979). Another argument was that the intellectual "class" in Hungary (and in eastern Europe) came to occupy a dominating position. They pretended to posses the monopoly of "teleological knowledge" which "legitimated the right of expropriation of surplus" (Szelényi and Konrád, 1979).

The intellectual opposition attempted to break with the taboos of official Soviet Marxism and also to warn against the dangers of a technocrat-driven, economic reform-oriented approach. Instead they advocated "humanistic Marxism". András Hegedüs, the "Lukacs disciples", the Praxis group from a political sociological point of view and János Kornai from a reform economic angle argued against central administrative control and allocation of goods and for increased autonomy for enterprises and commodity exchange. These arguments were highly innovative and were perceived as strongly oppositional. They were also welcomed by the critics, which we described earlier as socialist opponents, on the basis that they could help to provide more guarantees for the working class to challenge the expropriation of the surplus without changing the legal form of ownership. Economic reforms, it was argued, do not create revolution but challenge the dominant system of expropriation in state socialist economies. The view that by undertaking radical reforms and creating a specifically "socialist theory" to provide the theoretical basis for the changes, the regime could and should develop into an ideal "eastern European model" drawing from the negative experiences of both existing capitalism and state socialism, was shared by many in opposition from the early 1960s onwards (Szelényi, 1979).

Bugajski and Pollack (1989) characterised the Hungarian opposition in the 1970s and 1980s as a group of intellectuals left largely at liberty, experiencing only minimal harassment to ensure that their activities were strictly isolated from the working masses. This was to prevent any sizeable social movement which could develop into an uprising again, as in 1956. Hungarian dissidents themselves also maintained a relatively restrained approach to avoid provoking Moscow's anxiety resulting in a Czechoslovak type of oppression or the local leadership's willingness to continue the reform process. The Hungarian dissent, according to Konrád (1989), was there to be the opposition on behalf of democracy but at the same time they collaborated in securing for themselves a relatively trauma-free survival.

The most important aim of the intellectual opposition was to force the party leadership to advance reform programs. It was believed that by pushing ahead decentralization, and expanding market forces within the state economy and outside it, a process of embourgeoisement would occur, creating a middle class which would then counter the dictatorial bureaucracy and develop into a rational bureaucratic elite of professionals. The subject of reform had been widely discussed throughout the twenty years of economic reforms in the Hungarian literature. This reform process—which started off with liberalising regulations in the economic field in 1968 and (with some hiccoughs) continued as long as the regime lasted—was of consistent interest and worry. Concern that the reform process was not going ahead and had been halted worried writers all the time. The list of authors whose works appeared in state approved publications is very long: Berend, Fricz, Antal, Szalai, Nyers, Kulcsár, Giday, Nagy, Ferge, Szelényi, Kopátsy, Gombár, Lengyel and Hankiss to mention just a few. Their writings were often highly critical but were published in the politically "tolerated" category, although often provoking strong criticism from party officials and the sacking of editors as a result. Thus even legal publications fell sometimes into the category of opposition.

The Samizdat[3] was the forum for the illegal opposition. In their regular publications samizdat authors, such as Krassó, Vajda, Demszky, Rajk, Tamás Gaspar and many others published a variety of papers from concerns of the continuing reform process to objections of the political oppression and demands of political rights. The samizdat literature, such as Hirmondó, Beszélő and Magyar Füzetek, openly questioned taboo subjects, such as the one-party system and the Soviet influence, which were prohibited even in the most daring state publications.

The underlying reason for concern both among those inside and outside the samizdat was that reforms were seen as the only foreseeable political alternative to achieve a tolerable society within the Soviet bloc. The Stalinist experience made people try everything politically possible to escape the possibility of the return of that period and its methods but it was seen as a real threat. The unsuccessful "breakout-attempts" of 1956 in Hungary, 1968 in Czechoslovakia and that of the Solidarity movement in Poland, provided no reason to believe in alternatives other than gradual reform at best. Hungary, with its fairly successfully continuing reforms, became the only successful example of state socialism "with a human face", using Alexander Dubcek's words. The other existing political and economic examples within the Soviet bloc did not make Hungarians optimistic about available options. Dubcek's attempt in Czechoslovakia failed, the Polish experience did not create a better solution, the Yugoslav model also collapsed. Other Soviet-bloc societies, such as the Soviet Union, Romania or Albania showed the frightening options existing in the region. The prospects were extremely worrying and concern over the sustainability of reforms was felt very strongly. Samizdat and reform authors constantly kept up the debate providing advice on how to progress the reform process in order to prevent any U-turn. Even though Hungary was the most politically liberal society within the Soviet bloc, its political development remained in the unwritten, informal sphere, rather than in the realm of legal changes. Without establishing any legal and political institutions which could secure this political liberalism no-one could guarantee it if, either internal or external, political forces changed. The "Hungarian model", one can argue, went as far as a society could in political liberalization within the framework of a state-socialist system.

This informal political flexibility contributed to the development of a civil society with critical views. It did not, however, create a situation in which social movements could be sustained. The opposition was exhausted by the battle with the authorities to keep the Samizdat going, which was strictly illegal. Any serious attempt to organize a movement, however, failed at the start. This was the point about which the authorities were most "neurotic". Social movements dealing with any concern, such as peace, disarmament, or anti-nuclear issues, were crushed before they could develop. The threat that any potential movement would extend into an uprising, like that of 1956, was felt so strongly by the party leadership that it became a number one priority for internal intelligence to monitor and prevent it. An unspoken trade-off was offered to the people: as long as they did not attempt to organize revolution they could fulfil their need for ever increasing living standards and private property and enjoy political liberalism.

There were very few social movements which struggled against the odds. One of them, which was the target of constant police harassment, was *SZETA* (*Szegényeket Támogató Alap,* a foundation to aid the poor), which in a multi-party democracy would only be considered as a charity organization. Its aim was to collect second-hand clothes and distribute them to the poor, who were often Gypsies, the most deprived section of Hungarian society. But under state-socialism even this was considered to be a highly political and strictly illegal activity, and its main organizer

(Ottilia Solt) was subjected to regular harassment and atrocities from the police and was never allowed to travel abroad because of her "subversive" political behaviour.

In the period prior to 1988, when the regime had started to crack and movements had started to mushroom, there were only a few abortive attempts to organize any social movements. There was a case, when a few young people tried to gather support against both the American and Russian armies and put pressure on both governments to speed up disarmament. The hope was to increase opposition to both sides in order to emphasise the strong desire for peace and to reduce the risk of war by expanding or maintaining the arms programs in both the Soviet and American camp. It was hoped that the authorities would allow the creation of a peace movement with such modest aims since the official policy also proclaimed its "devotion to peace". After the first (and, as it turned out, last) peaceful demonstration, organized in a public park on Margaret Island (in the middle of Budapest) the police visited the core activists one by one and "explained" to them in a "friendly" threatening manner, that they had to stop the movement. The police argued that the existing official, state-maintained peace organization, the *Hungarian Peace Council*, (notorious for being a dumping place for retired hardliners), was there for any Hungarian citizen to join if they felt like "fighting" for peace. That was the end of the Anti-American and Anti-Soviet Independent Peace Movement of 1984.

Another peace movement, called Dialog, which started in 1982-1983 and died very shortly afterwards, was another exception to the rule that social movements were scarce in Hungary, along with the Danube Circle. The latter was an environmental movement, which was founded in 1985, and was the only social movement which continued after the regime change. (The Danube Circle will be discussed in chapter 5, along with other environmental movements.) The rapid failure of the early peace movement, Dialog, was due to a deep internal split amongst the activists at the first stage, from which it never recovered. "On the way towards becoming an organization, through the long series of discussions about electing a leadership, direct democratic, anarchist and representative democratic principles came into collision and shortly thereafter the 'radical autonomist' and 'moderate-constructivist' groups turned on each other" (Bozoki, 1988, p.388). They destroyed themselves before the police got to them. In 1982 in Hungary civil society was still underdeveloped compared with its later form by 1988. The lack of experience of these activists prevented them from developing enough maturity in their political behavior to cope with a situation like this.

The last example of movement activities before the change of government in 1989 was an anti-nuclear waste site movement in the village of Ófalu in Southern Hungary, which is located relatively close to the nuclear power station in Paks (Juhász, Vári and Tölgyesi, 1993). Ófalu was chosen as the "ideal" place for a nuclear waste dump. The decision about it was made behind closed doors. The first and only nuclear power station in Hungary started to operate in December 1981. Long before that, in 1976, a decision had been made to look for a suitable place for dumping nuclear waste, but it was only in 1983 that a location was found and accepted by the ministry, who then turned to the local authorities for formal permission. It was around 1987 when the village was informed by someone who lived in the village and worked at the power station about the plans for the dump.

The movement of the small village of 500 inhabitants against the giant of the nuclear power station, backed by the state, shook the public. The socialist government by this time (1988) was not strong enough to maintain its anti-democratic practices and the media was also allowed to report about it. The then opposition party, the freshly established Hungarian Democratic Forum, (between 1990-1994 in government) organized a national demonstration with well orchestrated press coverage. Other opposition parties also joined them. Many of the leaders of the movement joined the emerging political parties on both sides[4] of the emerging political spectrum. In January 1990 the Socialist government made one of its last decisions: the project was cancelled and the movement of the tiny provincial village had won a previously unimaginable battle against a state-supported giant. The leader of the movement, the chairman of the local council "won", too. He became a member of Parliament during the following general elections, in March 1990, representing SzDSz.

In summary, social movement activities were the thorn in the flesh for the Communist leadership in Hungary. This led to a situation in which in politically more liberal Hungary there were fewer social movements than in the politically more rigid Soviet Union. Hungary, however, was more tolerant towards publications, which resulted in a wide range of legal writings containing "constructive" criticism. The focus of this criticism was the nature and future of the continuing reforms which was also a central concern of the samizdat. Within the samizdat, however, topics were also discussed which were taboo in state owned publications, such as Soviet domination and the question of the one-party regime. The few Hungarian social movements which appeared before 1988 suffered, on the one hand, from police repression, and, on the other hand, from the activists' lack of experience in organizing a social movement and handling ideological conflicts. The only exception was the Danube Circle. In the period from 1988 onwards, before the collapse of the last Socialist government, a new, unimaginable wave of political activities including social movements started in the realm of civil society, which will be discussed in chapter 5.

Conclusion

Hungary and the Soviet Union shared a similar system of political institutions and a one-party system. The traditional "totalitarian" approach overlooked, however, that within this system there was a considerable number of people who opposed the regime in different ways. The "pluralist" literature and the literature on the opposition came closer to reality by recognising these groups and identifying the common characteristics among them in all socialist societies. The individual countries, however, show considerable differences, and Hungary and the Soviet Union provide a good example of contrasting concrete situations. In the Soviet Union there was less tolerance towards published criticism and fewer reforms were introduced before the Gorbachev period. In Hungary, on the other hand, the level of open oppositional debate was much higher in the publications but the number of organized movements was much lower than in the Soviet Union.

I now relate the discussion here to my earlier outline of the concept of civil society and my classification of the opposition. I defined civil society as those political activities which lie outside institutionalized state activities and party political activities, although the connections between civil society and the latter two types of activities are important. From the discussion above it is clear that civil society defined in this way started to emerge in the Soviet Union after Stalin's death and in Hungary after 1956.

In the Soviet case, under Khrushchev, it was encouraged and resulted in a variety of types of activism. Under Brezhnev many of these initiatives were repressed but this did not lead to their complete disappearance as under Stalin. This was, firstly, because of the accumulated experiences under Khrushchev and, secondly, because of the growing discontent under Brezhnev. In the short spell under Andropov, but to a fuller extent under Gorbachev, the expansion of civil society accelerated.

In Hungary after the watershed of 1956 a gradual process of opening up encouraged the development of civil society. The Soviet influence was significant in the sense that Khrushchev initiated the destalinization process. However, the Brezhnevian U-turn did not lead to a similar political reversal in Hungary. Gorbachev's new political approach, on the other hand, contributed fundamentally to the further development of civil society in Hungary and to the subsequent regime change.

Theory and Practice Combined

I now relate the opposition in Hungary and the Soviet Union to the earlier described conceptual discussion. In respect of the level of resentment, opposition in Hungary was at a higher level than in Russia:

there was a bigger core of radical opposition, there was more cynicism about the regime, and there was a much more active debate about reform. This was because of the higher degree of political tolerance allowed by the Kádár regime in Hungary and the greater exposure to western ideas compared with the Soviet case. The Hungarian opposition was stronger and the legitimacy of the regime was also more strongly challenged in Hungary than among Soviet people. This is related to two major factors. Firstly, the fact that Hungary had a much shorter history of Communism and secondly, that Hungarians felt that Communism was imposed on their country.

In Hungary the opposition which was most critical of the HSWP (Hungarian Socialist Workers' Party) took the form of extensive samizdat publication, underground lectures and a few clandestine social movements whereas in the Soviet Union there were more movements, a small number of well known dissidents, and less samizdat publishing or lecturing.

The majority of the population in both countries, however, while distant from Party officials, were not extreme in their opposition. In Hungary they were more questioning of the legitimacy of the socialist regime. In the Soviet Union, on the other hand, although discontent was present, there was a much stronger acceptance of the Soviet regime.

Finally, I come to opposition internal to the Party. In Hungary "reform Communists" were a very important category as they were in a position to publish widely and the political leadership was often willing to discuss their proposals. In Russia, on the other hand, their equivalents were far fewer because reform ideas were less tolerated (except under Khrushchev) and publication was more strongly controlled. There were also frequent debates in both countries among the Communist leadership resulting in opposition between the different factions. In Hungary these debates among the different factions within the top leadership were extensive though they never led to a change of leader. They did, however, modify the course of the reform process in both directions, creating the alteration between more progressive and "hiccough" periods. In the Soviet Union, however, opposing factions caused the downfall of leaders, such as Khrushchev. It also caused a very radical change of policy after Brezhnev. The latter was connected to the fact that under Brezhnev even internal Party debates were kept to a minimum level.

In terms of the intellectual/working class types of demands, as discussed earlier, there were no significant differences between the two countries. Similar methods were used (strikes, go slows versus abstract demands, such as freedom of speech and assembly) and similar contrasts existed: there was a gap between opposition by the two categories. However, nationalist and religious themes were more important in the Soviet "intellectual" opposition.

The overall result is that while the nature of opposition in the two countries was different (in Russia more social movements and less tolerance of reform publishing, in Hungary the reverse), the challenge to the legitimacy of the regime was much greater in Hungary. I now turn to examine environmental movements in the two countries after the regime change, first in Hungary and then in Russia.

Notes

1. The capital of Buryatia, seventy-five kilometres away from Lake Baikal.

2. The Állam Védelmi Hatóság (ÁVO) was the Hungarian equivalent of the Soviet NKDV, later called the KGB.

3. Samizdat is a Russian word by origin and literally means self-publication, referring to its origin as handwritten, illegally circulated political literature. The word was used for all illegally written and circulated political writings in eastern Europe, whether handwritten or not.

4. The two major political parties in 1990 were the Hungarian Democratic Forum (MDF) on the centre right and their major opposition, the Free Democrats (SzDSz).

5

Hungarian Environmental Movements: Case Studies and Analysis

In the previous chapter I looked at the theory of Soviet-type pluralism and civil society and the way opposition existed in the Soviet Union and Hungary prior to the regime change. I now turn to the case studies and the analysis of environmental movements in these two countries, based on our empirical studies in Hungary and Russia, conducted between 1990 and 1995. (Details concerning methodological aspects of the data gathering are discussed in the appendix.)

First, I will look at the case of Hungary and in separate chapters at Russia. chapter 9 will be a comparative analysis of the Russian and Hungarian cases, after which I will examine the relevance of western social movement theories in the eastern European context (chapter 10) before concluding the book.

The empirical basis of the book is the 129 interviews conducted firstly with environmental movement activists, including leaders of the groups and rank and file members, and secondly, with local authority members, including elected representatives and officials. The comparative analysis will build upon the four empirical chapters: chapters 5 and 6 will introduce several cases of Hungarian and Russian environmental movements. I will firstly describe their origins, aims and goals, participants, leadership, internal conflicts and conflicts outside the movement, their degree of success and the role of the media. In the second part of both chapters I will analyze the findings. In chapters 7 and 8, I will examine the movements' relations with the local and national authorities (in Hungary and Russia separately). These will be followed by chapter 9, the comparative analysis of Hungarian and Russian environmental movements focusing on the similarities and differences between them.

In the first part of this chapter I will examine the causes of environmental problems in Hungary and the growing awareness concerning the environment. I will then introduce several Hungarian environmental groups and finally I will analyze various environmental movements in order to evaluate the Hungarian green movement situation.

Introduction

In the one-party system interest and self-support groups existed but were only tolerated. They were kept under close observation by the authorities with the help of extensive networks of informers. The groups were judged individually and acquired a certain reputation in terms of their political radicalism which determined the authorities' degree of tolerance towards them. None of the groups could feel completely safe in the one-party regime as they had no legal protection. This constant potential threat obviously limited the number of groups as well as the number of participants, but did not succeed in completely eradicating peoples' willingness to organize themselves.

Social movements cannot be created by a favourable legal framework but, as political opportunity structure theorists would argue, it certainly encourages their existence. Consequently, we should not be surprised to see the mushrooming of collective activities which started as the existing legal constraints were relaxed just prior to the rapid political changes from state socialism to a liberal democratic system. In the second half of the 1980s thousands of groups appeared almost simultaneously (Bozóki, 1988; Juhász, Vári and Tölgyesi, 1993; Szirmai, 1997).

Industrialization in Eastern Europe and Environmental Problems

After the Communist Party established its power in Hungary in 1948, industrialization, and especially investment in heavy industry, was given absolute priority. There were two major reasons for this. Firstly, it was related to Marxist ideas which saw industrialization as the path for the development of a society, a priority which was similar to that pursued in other socialist societies. The other reason for pursuing industrialization was a much more practical one. The cold war period shifted the interests of the government in the Soviet bloc (as well as in NATO countries at this time) towards developing the economy to serve military aims.

Hungary, like all socialist societies, as well as western democracies, became part of this process which resulted in giving industry and especially certain sectors of industry, absolute priority in expenditure (Pető and Szakács, 1985). This industrialization policy, combined with the increased use of chemicals in the agriculture, created a severe environmental situation all over the developed world, and from in this respect rapidly modernizing eastern European societies were no exceptions.

Green Ideas in the Developed World

The recognition of the damage caused by industrialization and the excessive use of chemicals in agriculture started in the most developed societies in the 1970s. The first Green Party was established in New Zealand in 1972 and the first Green member of a national parliament was elected in Switzerland in 1979. Apart from pioneering exceptions the upsurge of green ideas did not appear before the late 1960s and early 1970s even in the most developed societies, where the success in raising living standards had previously silenced any criticism of industrialization. The first attempts to attack prevailing views appeared in Carson's *Silent Spring* (1965) on the use of DDT in agriculture, and the *The Costs of Economic Growth* (1967) by Mishan, followed by the establishment of the Club of Rome in 1968. Its report, *The Limits of Growth* (1972) questioned the ability of the planet's resources to meet contemporary rates of consumption. This was followed by *A Blueprint of Survival*, published by *The Ecologist* (1972), in which a need for national and international movements and a new philosophy of life was voiced (Richardson, 1995).

Green Ideas in the "Eastern" Bloc

Unlike the west, however, the eastern bloc did not suffer from consumption "fatigue" which could have led to the development of "postmaterialist" values. Although living standards had been improving dramatically in every state socialist society—relative to their own past standards—and, although the prevailing political regime emphasized this, the majority of people compared their living standards with "western", i.e. that of the developed countries' achievements. This comparison resulted in a desire to achieve a higher consumption level for people in eastern Europe rather than aiming at reducing it. As a result anti-consumption ideas did not develop in state socialist societies in the 1970s, as they had in central and northern European countries, such as Germany, Austria and Sweden, where living standards grew highest by the 1970s.

However, some state socialist countries, where living standards were highest in the region and the degree of political isolation was lower, were not immune to ideological influences arriving from the environmentally more advanced western Europe. When the above mentioned literature was published and news of the growing environmental awareness reached the most educated section of these societies, green ideas started to grow and a little later, by the late 1970s and early 1980s, environmentalist ideas had started to develop in Hungary as well. A survey conducted by Kulcsár and Dobossy in 1985 found that 80 per cent of the Hungarian population were aware of environmental problems and were very concerned about them

(Kulcsár and Dobossy, 1988). People were worried about air and water pollution, industrial and nuclear waste and the decreased extent and quality of forests. They also complained about the level of traffic noise and the decreasing proportion of green areas, especially in cities. There was also an awareness of the growing health hazards as a consequence of environmental problems.

Why were there no Environmental Movements before the mid 1980s in Hungary?

The development of environmental movements encountered serious obstacles. There were only three countries in eastern Europe where the organized opposition against the Communist Party became significant: Poland, with the Solidarity movement, Czechoslovakia, with Charter 77, and the GDR, with the pacifist movements organized by the Lutheran Churches (Bozóki, 1988; Dawisha, 1988; Bugajski and Pollack, 1989; Tismaneanu, 1990; Waller, 1992; Waller and Millard, 1992). In other societies, such as Romania, Albania and Bulgaria, party control was overwhelming and prevented opposition to the regime from developing (see Pickvance, 1998b).

In Hungary there was a one-party system but there was a lack of total party control. This allowed the development of a special type of resistance. Apart from a tiny group of dissidents and the occasional strikes among workers, the most developed form of resistance was constant, freely aired criticism, and a passive resistance, what Konrád (1989) called "antipolitics": a turning away from political questions and concentration on individual matters, personal careers, family life and raising individual living standards. The Kádár regime encouraged this individualistic response. The constant fear of "1956" being repeated made the political leadership feel safe as long as the population did not engage in independent civil initiatives. Organized opposition was a thorn in the flesh of the party leadership and it suppressed it even when, as from the late 1960s, political liberalism already tolerated individually expressed, "constructive" criticism.

Those few who preferred a less self-centred form of political resistance had to fight on two fronts: against party control, and against the political apathy and individualism which became widespread and characteristic in Hungary. In the above mentioned survey on environmentalism, conducted in 1985, respondents saw the solution to environmental problems either as by state action (tougher legislation, stricter control of polluting companies) or via individual action (paying more for better services to clean up the environment). The idea of organized action was scarcely mentioned by the people surveyed in 1985. It is not surprising therefore that the very few environmental movements which existed at this time were isolated and little known to the public, although environmental concern was present.

The Appearance of Environmental Movements

The first environmental movements appeared in Hungary from 1984. The earliest movements emerged simultaneously in large cities and small villages. Some were triggered by specific events, as in the case of the Danube Circle and the movement against the nuclear dump in Ófalu, in Baranya county, and others were reactions to a generally growing concern combined with the lack of any activity from the state bureaucracy (Reflex Movement). The sudden change in the willingness to form social movements was due to the changing political environment.

Thus political opposition was sporadic in Hungary prior to the mid 1980s, partly because of direct police repression, and partly as a result of the prevailing and successful "party line", which encouraged people to seek individual solutions to their problems rather than organized ones. When in 1987 more progressive Communist leaders replaced the older guard (Grósz), many previously "forbidden" names emerged in state approved periodicals (for example, Ivan Szelényi, George Konrád, János Kis, and András Hegedüs) discussing formerly tabu subjects.

In 1988 an interview with one of them, János Kis, who had formerly never been published in anything other than illegal samizdat publications, appeared in *Valóság*, a state-approved social science journal. In the interview Kis referred to the "democratic opposition", a term never before used outside the samizdat, and argued that from about 1987 the official authorities had started to remove constraints, step by step, allowing the dissidents to gain wider publicity and acceptance. In fact, party officials had even started to communicate with the "democratic opposition" in order to negotiate with them, and especially with those involved in publishing *samizdat* literature. At the time of the interview, in 1988, János Kis could not predict how long the process would take but he identified the starting year of the erosion of the existing socialist regime as 1987. The persecuted political dissident could sense the changing atmosphere in the first place, but the changing political circumstances were soon widely felt by the rest of the population as well because of the gradual opening up of state publications. These changes perceived by all led to the appearance of organized political actions.

In fact, the easing of political control led to a certain euphoria and resulted in the mushrooming of oppositional initiatives from 1988. The so-called "Round Table" negotiations started in Hungary which were initiated by the governing reform Communists. These included future political parties, as well as grassroots organisations of independent trade unions (TDDSz, Democratic League of Trade Unions, Workers' Solidarity, Union of Workers' Councils) and independent organisations of professionals (VOSz, the union of entrepreneurs, The Independent Forum of Lawyers, for example). Simultaneously many social movements were formed as well, like the Tenants' Organization, Homeless Movement, and most of the environmental movements, such as Green Future, Air Group, Fadrusz Street Movement.

Strictly speaking all these oppositional organizations were illegal at the time as the Bill legalising free associations in Hungary was only passed a year later, in 1989. The political and legal changes were brought about, on the one hand, by the progressive wing of the Communist leaders who were ready to share power and dismantle the one-party rule of the Hungarian Socialist Workers' Party and, on the other, by the pressure from the multiplying grassroots organizations "below" (Arató, 1992). In this period of regime change, for the first time since 1956, many people felt an overwhelming sensation of political freedom and hope. As one activist later, in the early 1990s, put it: "The first activity I have ever participated in was a demonstration in May 1988. Around 2-3000 people participated in this demonstration. It was very moving and exciting to me. I had never been involved in any political demonstrations or activities before" (Judit Kántor, a Danube Circle activist, p.34).

Environmental Movements in Hungary: Case Studies

In the early 1990s there were several hundred registered environmental groups in Hungary (Szirmai, 1997). They differed in size, locality and in their concerns which ranged from local issues to national or even global ones. It is not my aim to describe all these environmental groups here. Instead a few will be described in detail, illustrating different kinds of environmental groups in Hungary: a local movement which became well known all over Hungary, called Green Future; two different types of national movements, one of which became well known abroad (the Danube Circle), and another one which became very successful at the time when many thought environmental groups were on the decline; and finally an example of environmental movements outside the capital.

A Local Movement with a National Reputation:
the Case of Green Future

Some environmental movements concentrate only or primarily on local issues. The locality can be a provincial town or village or a particular district of the capital, as is the case of Green Future, which is located in the outskirts of Budapest, in an industrial area. However, as the example of Green Future will show, local movements can gain a reputation far beyond their localities.

Green Future started in the summer of 1989. The future activists of the movement became interested in green issues together with many other Hungarians who from the mid to late 1980s showed a growing interest in environmental issues, as a result of the upsurge in interest in Germany and Austria, which trickled over the border (Szabó, 1993).

The Origin of the Movement

Just before the birth of Green Future a local organizer of the community centre in Nagyteteny organized a series of lectures on environmentalism and invited several speakers on the subject.

> I was very naive and had little knowledge about environmentalism generally when I started to work here in 1987. I started to invite experts to give lectures for the public and the children: doctors, teachers, environmentalists came to speak to us. We all benefited from these enlightening talks. People became more aware of what was happening around us and started to be more and more irritated about the pollution in the district. (From the interview with one of the co-founders of Green Future, Ágnes Hársfalvi, p.1)

The lectures "opened people's eyes". They started to "see" the dark, black smog coming out of the local factory chimneys, argued Ágnes Hársfalvi (p.2). The lecturers pointed out also that these polluting state companies only paid negligible fines, which of course did not persuade them to alter their polluting behavior.

Local GPs in their lectures at the community centre spoke about their own observations, suggesting that local children suffered medically from the polluted air, soil and water in the district and that the level of cancer-related cases was higher in the district than the national average (interview with Dr Olga Kékessy, a local doctor, p.6). They suspected neighbouring factories, such as Chinoin Pharmaceutical and Metallochemia, as well as large pig farms polluting the river by discharging their waste into the Danube. The river was used for irrigation in the neighbouring gardens and allotments where vegetable and fruit were produced for home consumption. The local GPs drew up their own statistics based on their observations going back as far as 1977. These only became known locally via the lectures organized in the district community centre by a future movement activist in 1989 (Éva Utassy, Green Future activist, p.8).

There were other important local issues which mobilized people in the area. One of them was the plan for a ring road around Budapest, would be built right through a highly populated housing estate in the district. This plan had never been discussed in public meetings and was not known to the public until the construction almost reached their doorsteps. The main reason for building the ring road through highly populated areas, rather than a little further out of the city to avoid housing estates, was that a Soviet military base lay directly in the way of the ring road and, given the choice of disturbing the Soviet base or building a motorway through a highly populated area, it was obvious for the prevailing political regime what to choose. But, for very good reasons, they kept the plans very quiet. "The construction work and some documents leaked out only in the autumn of 1989." (Béla Sárossy, Green Future activist, p.14.).

Finally, it was very important that all these events occurred in the late 1980s, at the time when the "party-state regime"—as Ágnes Hársfalvi (p.18) put it—had its last period of existence and when the old political structure was starting to be dismantled.

Reasons for Joining

The hard core activists of the movement came together in the community centre around the lecture organizer.

"We were the first people who met in the lectures: Olga, a local GP, Béla, who was active in the ring road protest, Judit, who worked as an information officer in one of the local polluting companies, János, a graduate, Jóska, a former bus driver, later unemployed, who became very active politically, and the future local MP, Károly" (Éva Utassy, p.5).

The triggering event for forming the movement was that the local council had discussed the alarming environmental situation in the district in a meeting, but behind closed doors. (This was still the period of the previous regime.) The community organizer bravely decided to call a public meeting, inviting the local representatives of the just-forming "embryonic" parties in opposition. "I wanted the opposition to help us to put pressure on the authorities. These were the last days of the party-state power,"she said (Ágnes Hársfalvi, p.3).

The future core of the leadership did not know each other very well previously and were occasionally hostile to each other at the beginning. Not even the brave community organizer escaped this initial attitude (community centres were run by the local councils). "Many people were hostile to me at the beginning. They just could not believe that someone employed by the local authorities could be trustworthy. But I proved to them that I was" (Ágnes Hársfalvi, Green Future leader, p.4).

Three members of the future leadership called another public meeting a few months later, in September 1989, with the specific aim of going public and recruiting activists for the "District Environmental Council" as the movement was then called. "We wanted to emphasise that we want to represent the interest of the whole population [in the district] and attract anyone interested to join" (Béla Sárossy, p.12.) Forty people joined the movement at this public meeting.

The Movement Participants

Apart from the core eight to ten members, most of whom were the leaders, there were sixty to one hundred activists "available for whatever we asked them to do" (Dr Olga Kékessy, p.9) and a further 400 people regularly turning up at public meetings, demonstrations, and signing petitions. The activists considered sixty an ideal number in terms of organization and did not wish to increase this number. The circle of sympathizers was wider still but the precise number was unknown to the movement activists as they have never had the means of conducting a survey about it. They claimed it was huge, which was demonstrated during the 1990 election campaigns when the most important local issue was environmentalism. They felt well supported by the local population during the campaign.

The sixty or more activists who regularly paid their membership fees had mainly medium level education [A-level equivalent] but many had less. "We are popular right across the board in terms of social and educational background. There is not much conflict regarding educational differences" (Hársfalvi, p.8.). The majority of the activists were middle aged or older and women, often mothers with children. Women—the local activists believed—were more sensitive to environmental issues and more willing to do something about it. Often the whole family joined.

The Leadership

The leadership itself was 90 per cent made up of professionals except for, Jozsef, the unemployed bus driver. They were biologists, medical doctors, engineers and scientists. The only profession they felt short of was lawyers, who would be able to help the movement to deal with bureaucratic issues, legal requirements and illegal acts by the authorities. Green Future's leadership was a fairly rare type. It was fully democratic in every sense. They did not maintain a strict hierarchy even though there was a president. His role was only formal, he was the local MP and was kept busy in his parliamentary job. The rest of the seven to nine people shared the different kinds of task. This, however, did not mean that they did not experience conflicts and serious debates in the course of their activities.

Conflicts

Disagreements grew over a fairly lengthy period. One source of the disagreements stemmed from the diverse party political affiliations which developed following the phase when opposition parties established themselves before the national elections of March 1990 and six months later during the local election campaign. Some felt closer to the MDF advocating nationalist-Christian values while others felt nearer to the more cosmopolitan, liberal oriented Free Democrats or the young democrats, the FIDESZ. At this time there was no significant sympathy towards the Socialist Party. (However, by May 1994 the Socialists became the most popular political party in Hungary and won the national elections with a landslide victory.)

Another related source of conflict was that while the district voted for an MDF Member of Parliament in March 1990, half a year later, by the time of the local elections, a Free Democrat local government was voted in. This was similar to the national trend. The movement leadership became divided over the question of cooperation versus confrontation with the local as well as national government. These divisions were strongly influenced by their individual political affiliation with the different political parties holding power at various levels of authority.

Finally, the movement activists ran into conflicts over finances. This conflict was a product of their increasing success: the more funds they managed to attract the more money there was to row about. They started to accuse each other of mishandling finances and accounting and creating full time jobs for themselves. "Money caused more problems among us than the lack of it. When I was in charge of the accounting Éva started hostile rumours in the group that I was mishandling the money. It hurt me because it was unfair. I have never done anything like that" (Hársfalvi, p.14.).

The Goals of the Movement

Green Future is an example of a social movement which was originally organized with very concrete aims. They were twofold. The first aim was to fight the planned road project and to try to divert the route from the highly populated estate. The second concrete issue was the neighbourhood's largest polluting factory, Metallochemia. The activists and supporters strongly suspected that the main source of the diseases in the district was the negligence of the large chemical company which consistently mishandled chemical waste, dumping it on the company site, which was in the middle of a densely populated residential area. The movement's aim was to press for a government enquiry which would prove the existence of a dangerous situation and lead to the closure of the company, financial compensation and the cleaning up of the site.

But at a later stage, after their first successes, the movement developed plans of a more continuous nature. They decided to monitor and continuously measure the level and detailed content of the air, soil and water pollution in the district as well as collecting systematic statistics of the diseases and general health status of the population of the district. They knew that these were the tasks of state organizations, such as public health authorities, but they had no trust in them. The movement decided to take on the role of "representing" the district's interest in all environmentally related questions. One reason for this was that the movement activists felt they had developed expertise on environmental issues during the initial period of concrete fights. This would have been wasted unless turned into a more systematic and ongoing activity. They also felt that, on the one hand, they had become better equipped to fight with the local and national government but, on the other hand, the new regime had only changed in legal terms. It had become democratic, tolerating social movements and different political parties, but it was felt that the regime remained as resistant to public pressure as before. The new national and local governments did not improve their reputation in terms of introducing radical changes to solve the burning environmental problems. The need for ongoing pressure was felt very strongly and not only from the activists' side. The population of the district started to turn to the movement with countless requests, partly acknowledging their success and experiences as well as their growing expertise in handling and evaluating cases, and partly in the hope that they could or were more likely to be able to get results than private, individual actions. Finally, and perhaps most importantly, the movement developed among its goals the "education" of the population in a very wide sense. They started organising lectures again, pursued recycling, and paid special attention to educating the children by organising special environmentalist summer camps for them.

> We wanted to achieve more than just concrete goals. We wanted to educate people and to try to achieve preventative measures concerning the environment, not only cleaning up the damage afterwards. For example, we went out to the nearby forest of Haros and listed the zoological and botanical importance of that ancient forest in order to get official recognition of its importance in the future. We also organized camps there for young people and lectures for adults in order to educate them. (Interview with Lajos Salgó, a Green Future activist, p.5).

Success

Success can be measured in two ways: firstly, tangible success and secondly, long-term intangible success. Examining firstly the tangible type of success, Green Future became successful in one major concrete area. They managed to have Metallochemia, a large company of national importance, closed down. The government inquiry concluded that their strong suspicion was justified and had to act upon it without delay. Plans for cleaning up the site were drawn up and talks about compensation started. The government even allocated a special environmental fund for the district to improve its pollution overall. The question of the ring road, however, had not been resolved. The construction of the part outside the district developed to the point where the diversion of routes would be a lot more difficult. The movement activists, however, have decided to continue their fight regardless.

When asked about their success the movement members considered people's changed perceptions (the second type of success) to be their most important achievement. First, the population in the district, they argued, look at nature, the environment and waste differently from the period prior to their activities. The second most important intangible success was that people in the district had learned to represent their own interest. They cannot, as in the past, be excluded from information and decision-making. A road project for example could never now get to the complicated, entangled stage before protest or consultation could start. And finally, the movement achieved not only a cognitive acknowledgement of environmentalism but a political one as well. "No political party can even imagine being elected in this district without being interested in environmental questions," they argued (Judit Kovács, Green Future activist, p.16).

Media

Part of Green Future's success is due to the local and national media. The formation of the movement coincided with the newly found "free" period when the traditional state control of the press had disappeared. Although, between 1990 and 1994, under the conservative right wing MDF government, some degree of "censorship" had been re-established, this did not concern environmental matters but other political issues. Similarly, this was the period of environmental "awakening" nationally, as well as locally, and so information about a well organized environmental movement was welcomed by newspaper, radio and television editors.

The media is very important. Sometimes they misinterpret what we say, so I cannot say I trust journalists, but we badly need publicity. We are very often mentioned by the media: in papers, by the radio and TV. Apart from that, I frequently appear also on the local cable channel. Last time, for example, I talked about my new refuse collection plan. (József Hollán, Green Future activist, p.7)

Green Future became one of the environmental movements which became well known. It was founded during the regime change and was consequently lucky enough to get all the publicity it needed in the national press. Its fairly quick and relatively spectacular success earned it yet more publicity and national recognition. It could also be argued that this national fame further contributed to its own success. The surrounding publicity increased the pressure on the government to act and the case of Metallochemia was felt to be sufficiently serious for the government to feel it had to do something about it to avoid further embarrassment. The unexpected financial "windfall", the setting up of the special government fund to aid the district's environmental development, strongly supports this argument.

Summarising the case of Green Future, it is a good example of a generally fairly successful local environmental movement which achieved national fame. It has expanded its original goals which were concrete and well defined. The newly developed long term aims concerning "public education' both about environmental issues and political actions were similarly clear and were based on their own enlarged capacity to tackle environmental problems to handle the press, national and local authorities, and to achieve their goals. In turn, both the publicity and their success contributed to their national reputation. More importantly, they have earned the respect and trust of the local population, a factor which strongly contributes to their survival.

National Movements

I will now examine two national movements which represent two contrasting cases. One—the Danube Circle—is an example of a movement which has been around for a long time. It was very popular at a certain stage after which its popularity plummeted, though the movement itself has survived this. The other example is the Air Group, an originally relatively small movement which has catapulted to national fame and made it much more important than it ever planned to become.

The Danube Circle

The Danube Circle is probably the only Hungarian environmental movement which has achieved international fame (Waller, 1992; Fleischer, 1993).

The Origin of the Movement

The idea of a dam jointly built by Hungary and Czechoslovakia at Bős-Nagymaros was first raised in the 1950s. The power station was supposed to supply energy, help navigation and control floods. The agreement was signed by the two countries much later, in September 1977. The plan was to build two power stations, (one at Bős/Gabcikovo and another at Nagymaros) and a twenty kilometre long reservoir. The dam was supposed to ensure a four to five metre flow wave passing several times a day. The monstrous plan has all the fingerprints of the engineering ideas of the 1950s and 1960s when such constructions were built all over the world, and especially in the less developed part of the world, often with first world "aid". The argument was always that they would provide low cost energy. Apart from the obvious environmental damage in most of these cases, even the "low cost" argument ceased to apply by the 1970s. In the Bős-Nagymaros case, for example, the very expensive investment would only produce an insignificant amount of energy, about 2-3 per cent of the required amount in both countries, and the problems of navigation and flood control could have been solved by other, cheaper, means. The most important criticism of this giant plan, however, concerned its environmental effects. The dam and reservoir system threatened the drinking and underground water supply and the livelihood of the neighbouring natural habitat and the river itself stretching from the Austrian border to the middle of Hungary. Similar plans for a dam were floated in Austria at around the same time as the one on the border of the two socialist countries, but a national referendum swiftly rejected them. No referendum or public discussions took place in either state socialist Czechoslovakia or Hungary (Szirmai, 1997).

In fact, what happened was exactly the opposite. The plans were kept secret for a long time. The first limited public debates occurred only as late as May 1980 at a conference of engineers, which soon led to further debates, even if still mainly among people connected with the subject professionally (Fleischer, 1993). The 400 engineers who participated in the meeting in the House of Technology voted against the project as soon as they heard about it.

Public debates, however, started only later, with the intervention of János Varga, who later became the leader of the Danube Circle. Varga was not an engineer, but was a biologist by profession, a journalist on an environmental-oriented magazine, when he stumbled into the subject. Parallel to this, several local authority committees of the region questioned the feasibility of the project. Several national institutes, such as the

Hungarian Association of Hydrologists, the Union of Engineers and Natural Scientists and the National Office of the Protection of Nature and Environment joined the debate. Finally, the Academy of Sciences, as the most prestigious scientific institution, was asked to comment on the case. The special committee of the Academy recommended the abandonment of the project or at least, if this was politically too sensitive, the alteration or delay of the existing contract with Czechoslovakia, to allow further investigations (Fleischer, 1993).

The clear message from many sections of the profession did not lead to the logical reaction. Instead the political leadership classified all the documents on the subject as confidential and secretly gave the go ahead to the project. The wider public was still unaware of the storm in professional circles, and the media was not allowed to report about it. Not until 1984 did it become known to anyone outside the profession when a small circle of dissidents attended a meeting where János Varga explained the case. Many of the professionals who had been at the meeting in the House of Technology were also present. The "dissident" meeting turned into a movement: those present voted for a resolution to organize a group in order to raise public awareness and gain public support, put political pressure on the government and try to stop the construction.Reasons for joining

Many professionals joined the movement, including people who were aware of the ongoing debate and felt let down when their strong recommendation against it was ignored by the political leadership. But not only professionals were present at the meeting or joined the Danube Circle later. Some people joined because they had strong feelings towards nature generally and the Danube in particular. "In 1979-80 I heard about the dam plans and was outraged, this was why I wanted to join the Danube Circle" (Kálmán Kemény, Danube Circle activist, p.2). Others supported the cause following their awakening interest in environmentalism.

The Danube Circle became the first national movement with wide public support, the strongest public opposition against the government's ways of conducting important decisions. It became the only social movement well before the regime change of 1990 and attracted millions of sympathizers, signatures on petitions, and tens of thousands of demonstrators in front of the Parliament building. It actually reached its peak of popularity before the new regime had been established unlike any other social movement in Hungary.

I have been an activist of the Danube Circle since the beginning of 1980s. I used to help to collect signatures for petitions and joined demonstrations. There was a major demonstration in 1987 in front of the Ministry of Environment including Austrian Green participants. The Police brutally beat up people, Hungarians and Austrians alike. After that there were demonstrations at Nagymaros and in front of the Parliament building. Finally the Nemeth government [1988] promised to stop the construction. Our movement was strong and influential before the regime changed. (Irén Szalai, Danube Circle activist, pp.1-2)

The Movement Participants

At the original stage, in the mid 1980s, the movement participants were mostly people who were part of the so-called dissident circle. They were often people with excellent academic records who had politically distanced themselves from the establishment and who were not prepared to embark on a career a within it. An alternative route was to become part of the growing group of the so-called dissidents. The circle of dissidents was amorphous and undefined. Apart from the hard core members who were engaged in writing, editing and publishing samizdat literature, anybody could be labelled as dissident or see themselves as a dissident if they belonged to a certain group of people who discussed political subjects on a regular, organized, club type basis and/or refused to advance their career by joining the establishment. Most of those people who joined the original Danube Circle were dissidents in these terms.

Later, in the late 1980s, the situation changed. Many former dissidents became leading politicians. However, the Danube Circle still attracted mainly well educated people as core members. Of course the demonstrations or petitions, which at their peak mobilised 40,000 people, attracted a wide spectrum of the population.

The Leadership

The leadership, just like the core members, consisted of educated people, biologists, economists, engineers, often with experience in academic work or publishing. The leadership, unlike Green Future's, was highly structured and strictly hierarchical. Of the top three, the main leader of the movement was János Varga. He was undoubtedly the most important and dominant character within the Danube Circle. He provoked strong emotional reaction among all participants either in positive or in negative terms. He was obviously a good example of a charismatic leader, with enormous intellectual appeal to most, but with a fairly low tolerance level towards those who disagreed with him. Several crisis situations occurred during the history of the movement, always concentrating on Varga's personality and ending up with people, or groups of people, walking out of the movement as a result of clashes and heated debates. However, those who stuck with him and accepted his leadership style felt equally strongly and positively when talking about him. These people were attracted to his intellectual ability, innovative views and charisma. These movement members found him irreplaceable and were fearful of losing him whenever a crisis occurred and he threatened to resign.

Conflicts

Most of the conflicts in the Danube Circle occurred as a result of the above mentioned personality clashes between Varga and one or more members of the movement. As Varga was the founder of Danube Circle, and had a fairly large group of supporters who remained loyal to him, the conflicts always ended with the other person, or persons, leaving the group.

> János Varga is a very special person. He is the core and the soul of this movement. I wish there were many others like him. He is a very rare person, a person of strong principles, and he cannot be corrupted. He is also very good at understanding problems which could threaten the movement, and solving them. But there was a time when he was not appreciated and wanted to leave the movement. He actually left the movement for a while. There was a huge publicity following that. I remained the only contact person between Varga and the Danube Circle because he refused to talk to anyone else from the movement. But luckily we solved all these. He returned to the movement and is with us again. It was just a short spell. It is over. I am very happy this way. I would not want to imagine this work, this movement without him. (Júlia Fejtő, Danube Circle activist, p.4)

As Varga has developed an enormous reputation outside the movement as well, he was courted by politicians to join, or at least openly support them, invitations which he consistently turned down. He maintained political "neutrality" as far as political parties went, even if he was viewed as a sympathizer of the Free Democrats. He also developed an international reputation, which led to the award to the Danube Circle of the so-called "Alternative Nobel Prize" and membership in countless international organizations, accompanied by the interest of the western press and very generous western funding.

The Goals of the Movement

The original aim of the Danube Circle was obvious: to stop the construction of the dam at Bös-Nagymaros. But, like many other environmental movements, it has expanded its concerns and become a centre with an interest in a wider circle of environmental problems. Whether it has achieved success or not is more difficult to assess. It secured the decision by the Hungarian government to halt the construction on the dam on the Hungarian side. As result the two governments (Czechoslovakia at the time, later Slovakia and Hungary) locked themselves into never ending arguments and re-opened negotiations in search of new alternatives. Finally, a new plan, the so-called C-variant, of the dam has been completed by the Slovak side, while Hungary has not built the originally planned part, influenced by the strong public pressure against it. The fear, which the Danube Circle argued so strongly, was that the dam was damaging for the environment. This decision attracted western media interest, but also led to a court case at the European Court in The Hague, as a result of the sharp conflict and political disagreement between the two countries.

The Danube Circle became a political force with a considerable impact on both the Hungarian government's decision and on public opinion. However, it became somewhat the victim of its own success. It became such a well known movement, with such an enormous circle of sympathizers, that it could obviously not maintain this level of "revolutionary" popularity over the long term. The movement members discussed the possibility of becoming a permanent political party but this was rejected. Consequently when the political turmoil settled within the country, and demonstrations as vehicles of political pressure transformed into different forms of political activity, most people started to perceive the Danube Circle as a movement which had reached its peak and was on the decline.

In sum, the Danube Circle undoubtedly became Hungary's best supported environmental movement at its peak time. But (a) many people demonstrated with it at the time when demonstrations were a highly popular form of political expression, (b) it fought for a concrete goal of national importance which was easily identifiable for a large section of the population whether or not they had any other type of environmental interest; and (c) at the time the Danube Circle inevitably took on another role as well, which can be described as that of being *the* opposition. Any independent movement could become a symbolic vehicle of political opposition feelings towards the state socialist regime. The Danube Circle certainly performed this role at the time of its foundation and during the first period after the regime change.

In view of the above conclusions what is surprising is that the Danube Circle still exists. There are two reasons for that. The main reason for its continued existence, as in the case of Green Future, is that it has transformed into a green movement with a wide range of environmental concerns, beyond the hydroelectric dam on the river Danube dispute. As the Danube Circle's fame grew its political influence also grew, as a potentially powerful pressure group, with significant political influence. Thus the Danube Circle became a vehicle for pursuing various green matters. The other reason lay in the "human factor". The activists of the Danube Circle enjoyed participating in the movement, and were happy to devote their spare time to it. Being a member of a nationally and internationally known social movement gave considerable pride to the members. Many activists felt that they have gained experince in organizational skills and had to continue with the hard work. They could not give it up even when the group had difficulties or when it lost some of the mass support.

The Air Group

The second example of national movement is the Air Group. It certainly differs from the case of Danube Circle in that it was not around as early as the Danube Circle and did not have the chance to become a "dissident movement" in the same way as the Danube Circle. The Air Group achieved its biggest success at the time when many students of social movements studying eastern Europe were already talking about the disappearance of environmental movements in eastern Europe.

The Origin of the Movement

The Air Group came into existence during the turbulent years of 1988-1990. Prior to its formation there were three clubs, two of which were university clubs, (the Green Club of the Technical University and the Environmental Club of the Eotvos Lorand University plus the Environmental Club of the Esperanto Union) which can be viewed as the predecessors of the present movement. The Air Group's founders came together as activists in these clubs and decided to found the Air Group.

Reasons for Joining

The people who joined the Air Group felt sympathetic to the concern of the founding members about the high concentration of pollutants in the air, especially in major cities, and in the country generally. The movement participants were, again, mostly highly educated people. This was partly because its predecessors were university clubs but partly because this movement too was happier to attract people who were specialists in the analysis of air pollution or health-related problems. As the movement grew, however, so did the range of its participants and it started to accommodate a wide section of the population regardless of their education level. The core, though, remained well educated.

I am a founding member of the Air Group. First we had mainly students among us. Our love towards nature was the basis of our commitment to the group. Then we turned our attention to the problems of air pollution in the cities. We did measurements in the university labs as a kind of practical. Later we decided to organize ourselves as a social movement. We became more political: started to lobby and petition the government. This was from 1988. (Andrea Nagy, Air group activist, pp.2-3).

The Leadership

The leader of the Air Group is a geophysicist. He could not be more dissimilar in character from János Varga. He, too, was a charismatic leader, and was well respected by the group members, but has a much more peaceful character and a much calmer style of leadership.

The Goals of the Movement

The Air Group came into existence, unlike the two movements described earlier, with two aims right from the beginning: it had both a concrete goal and a wider agenda. The concrete goal was to achieve cleaner air by reducing private traffic, improving the quality of public buses and providing proper facilities for cycling. The wider aim was to combine environmental forces by attracting existing members of environmental groups scattered around the country and the city who were acting fairly independently from each other. These aims were taken right from the very beginning as conscious objectives, unlike the case of the other two movements where the idea of widening the movement's concern from a concrete goal to a more embracing general environmentalist approach developed at a later stage, almost as a survival tactic.

Success

The Air Group achieved an unprecedented popularity way beyond its circle of sympathizers at time when many environmental movements had already achieved their concrete, short-term goals and had arrived at their "second stage", when they started to transform into movements concerned with more general environmental interests, in order to survive. The Air Group managed to gain popularity at a time when the political excitement of the very first years had already calmed down. Its popularity is surprising considering that neither of its aims suggest any spectacular attraction beyond the small circle of the environmentally active section of the population. Nevertheless, the Air Group managed to become one of the best known and most popular environmental movements in Hungary. It came into the centre of interest well after the decline of the Danube Circle.

The reasons for its popularity differ fundamentally from those of the Danube Circle. It is not engaged in "opposition" roles on the political platform, nor does it pursue particular national interests. The reason for its popularity lies in the fact that it has struck a chord with the growing environmental awareness in the Hungarian population. The many environmental groups which paved the way for the Air Group have educated people and consequently have achieved one of their most ambitious aims: the raising of environmental awareness among the population. When the Air Group started to publicise their demands they were addressing a public which was already much more open to the subject than their predecessors' "audience".

The Air Group made people suddenly feel that they were indeed suffocating in the highly polluted cities and many joined them and volunteered monitoring the air in urban areas. When this was achieved and the measurements were made public, people's opinion became very supportive, helping the Air Group to pursue its aims. It managed to achieve an unprecedented governmental decision: a special environmental tax was introduced, on top of the normal state tax, on every litre of petrol, exclusively to finance environmental projects. The Air Group's aims were (a) to penalise road users for not choosing alternative means of transport, (b) to reduce car traffic by raising petrol prices and hence ease one of the main sources of air pollution in urban areas and (c) to create a special fund for environmental purposes. This tax was introduced with wide public support, instead of resentment, as is usual for new taxes. Further taxes were pressed for by the Air Group to penalise other road users, such as trucks. Although the petrol tax hit the population directly, it was accepted as a useful government measure, as a result of the Air Group's popularity.

> I consider the biggest of our successes that people started to think differently about the environment and the issues of pollution. Secondly, that we achieved changes in the legislation. The environmental tax is a tremendous success but there is a lot more to do. We regard every step as a success and see our task as a continuous long term duty. (József Mizsei, Air Group activist, p.19)

The Media

Publicity played a vital role in the movement's rapidly growing popularity. The Air Group activists were clever enough to turn to companies whose interest coincided with their efforts such as, for example, the biggest public transport company running Budapest's transport services. These companies agreed to finance leaflets and fliers and place them on their own boards inside and outside the buses, trams, etc. publicising the Air Group's recommendation that people should use more public transport facilities and leave their cars at home. The publicity campaign made the Group known everywhere and their innovative ideas gained the media's support as well.

They also publish their own magazine, *Breath*. As their publicity grew, so did the number of sponsors, and the strength of these made them more influential in their parliamentary lobbying.

To sum up, the Air Group's success in the "second" stage of development of social movements, following the immediate period of the regime-change when most social movements were founded in Hungary, is clearly the result of its innovative way of adapting to new circumstances. They found a new approach for an eastern European movement, persuading an enterprise to advertise their messages. Thus their environmental interest was combined with business interest to achieve fresher air in a suffocating city by encouraging the usage of public transport instead of private cars. They also recognised right from the beginning that short and long term aims can and should be combined in an environmental movement. Finally, they became popular at a time when Hungarians were becoming more aware of environmental problems, and becoming more responsive to social movements' persuasion.

A Movement Outside the Capital: Reflex

Finally, we consider an environmental movement outside Budapest. With its two million inhabitants the Hungarian capital is disproportionate in size to the total population of ten million, and like many capitals, plays a decisive role in the life of the country. It is also the capital of a society which was overcentralized in its socialist period. The result is that many national organizations end up being located in Budapest and develop a Budapest-centred view. As, however, 80 per cent of the Hungarian population lives outside the capital we cannot only focus on environmental movements which operate in the capital. According to the literature numerous environmental groups exist in the provinces (Juhász, Vári and Tölgyesi, 1993). To pay attention to environmental movements outside Budapest is therefore crucial.

Reflex is a movement which is located in Gyor, one of the five largest cities in the country, in the north western part of Hungary. It is a city with a strong tradition of being a cultural centre, as well as being an important industrial centre. Geographically speaking, it is closer to the Austrian border than to Budapest.

The Origin of the Movement

Reflex is actually almost as old as the Danube Circle, as it started its activities in 1985. But it differs from the latter in that it was not organized around any concrete objective at the time of its foundation. It came into existence by the decision of eleven people, who knew each other previously, and shared a strong environmental concern. Many of the founding members were working in the Environmental Office of the local authority at the time and felt frustrated by the lack of possible action within the framework of an "over-bureaucratized" state office, as they put it, where no actual work was demanded from them. Most of them were in their early thirties and were educated as biologists, environmental engineers, chemists and hydrologists.

The Movement Participants

There were 400-500 members who formally registered with the movement and regularly paid their membership fees and forty to fifty of them were activists who participated in the movement on a regular basis. The circle of sympathizers, who regularly signed petitions for them, was large, around 10,000 people. Considering that the whole town only has population of 300,000 this is proportionately large circle.

Half of the movement participants were young, below twenty-five, and both sexes were equally represented among them. But the leadership was predominantly male, there was only one female leader. Reflex paid a lot of attention to activities outside the city, in the rural areas and small villages, where environmentalism was far less developed than in urban middle class areas and where the population was usually older. They gained a considerable reputation in the region and were regularly invited to village meetings to participate in debates regarding local issues and to highlight environmental aspects. Some issues occasionally created difficult dilemmas between them and the locals, as when they advised them against new road being built between two villages through a nature reserve, while the locals preferred, perhaps understandably, to concentrate on the advantages of a new road. On the whole though, they were successful in recruiting people or winning their sympathy.

The Leadership

The leadership consisted of seven members, one of whom was elected as president. They were mainly the original core members, though some have left to pursue political careers. The president has been the same person since 1987, except for one year, when someone else was elected for the job. Elections took place yearly. The leadership was remarkably friendly with each other. They have been working together for a number of years now, surviving crucial regime changes, and yet they have only experienced few internal conlicts. But they were predominantly male. There was only one female leader although half of the activists were women. Reflex was different from other environmental groups in this respect.

Conflicts

The peaceful and constructive reputation within the movement did not mean that it had no enemies. The local government has had only one consistent characteristic through the crucial regime change: to remain consistently antagonistic and hostile towards the Reflex movement. They viewed them as potential enemies on many questions and developed a competitive attitude towards them. This is not typical in the Hungarian context where local authorities are usually responsive and offer some sort of cooperation with the environmental movements. (See chapter 7, which discusses the role and behavior of local and national authorities towards social movements.) The local authority, like the national government, went through radical political changes during this period. But its antagonistic attitude remained unchanged.

However, not all "official" bodies were, or are, antagonistic towards the movement. In the Communist period as well, as since, there were organizations which offered cooperation with them. In the 1980s it was the Young Communist League (KISZ) and the Hazafias Népfront (National Popular Front) which expressed their readiness to embrace them and in the 1990s the newly born political parties wished to be associated with Reflex, because of its good reputation in the region. But all these approaches were turned down by the movement activists and there was a conscious effort to maintain political independence and a global environmental interest. Among those turned down was the Green Party itself. This is because the movement maintained strong "anti political party" views, and wanted to keep the movement outside the party political framework rather than affiliate with any political party including the Greens. Individual members of the movement did of course have individual preferences towards one or other main political party. They even advocated accepting anybody's membership independent of their personal political views. The idea of political "independence" only referred to the movement as a whole.

The Goals of the Movement

Reflex's primary focus is global environmental education.

> Our actions are mainly to draw people's attention to certain environmental problems and to make people aware of environmental dangers. We organized a day of the "Earth and People" and other similar events. We tried to explain to people where the economy and political bias led us in terms of the environment. I value most our work educating young people. We have done that ever since we started this movement. We put a lot of emphasis on educating people by using posters, organising children's clubs, street demonstrations and many other methods. (Bálint Csaba, Reflex activist, p. 4)

The movement has many young participants because they concentrate on educating young people by frequently lecturing in schools. Students and secondary school pupils started to develop an environmental interest after coming across Reflex's lectures and developing the relationship into activism through these channels. The fact that many Reflex leaders themselves were fairly young (in their early thirties) helped them appeal to the younger generation. Another important factor was that many of them were either former teachers themselves or had a diploma in education. Interestingly, however, Reflex also developed concrete demands at a later stage of its activities. But, unlike other movements, Reflex had started off with strong abstract views and developed concrete objectives later.

Success

One of these concrete objectives has been achieved: Reflex suggested a ban on all cars in the central shopping area of Gyor city and has achieved it. The pedestrianization of the city centre created a much friendlier area where people could walk, meet, sit down or shop without being disturbed by the traffic, noise or air pollution. The other successful concrete outcome of Reflex's activities is related to road traffic diversion. A new, major motorway construction was taking place outside the city for which all road traffic, including construction traffic, went through the city. They managed to impose a ban on trucks (three tonnes and larger) driving *through* the city. They were also successful in persuading the city to improve recycling and at the same time in educating the population to use the facility.

Where neither they nor other social movements achieved full success was the Danube dam question. As the dam is actually located in their region, Reflex obviously took sides in the matter, strongly supporting the Danube Circle. It is interesting to point out here that (a) Reflex was not founded because of the Danube dam project. At the time when Reflex was founded in 1985-86 they hardly knew about it, given the lack of information available to the wider public outside the profession and "dissident circles", (b) they had never "claimed" any special rights over the Dam issue, even though it was geographically very close to them and would damage their region, and (c) they had never felt any competitive or antagonistic feelings towards the Danube Circle, which gained all the national publicity and fame over the matter at that time. Interestingly, Reflex is currently a stronger and more popular environmental movement than ever, while the Danube Circle has declined. But at the time Reflex not only did not get fame but even jeopardised itself locally by provoking the local government, which disliked Reflex's open and unconditional support for the Danube Circle and the fact that it went to demonstrate with them side by side.

Media

The local authority was very keen on influencing the local press against Reflex. They have achieved this through one of the local papers, *Kisalfold* (the largest local newspaper, named after the region), which ran a series of articles openly attacking Reflex's attitude on the Danube question, and supporting the local council's view which favoured the construction of the dam. But this paper was not the only local medium and national papers were also very favourable towards the movement as well as the part of the local media which was not influenced by the local authority. The Reflex group frequently appeared in the most important national papers as well as television (local and national) and radio.

> The contact with the media is extremely important because the role of the media is to inform the wider population. And the media is the best tool for us to reach the widest circle of people. We have never had any trouble contacting the press, the radio or TV. More and more frequently the journalists themselves seek to get information from us in order to report about us. Thus most of the time our relationship with them is very positive. (Gábor Szűcs, Reflex activist, p.10).

In fact, their fame went beyond than the country's boundaries. Reflex developed good contacts with several international organizations in Austria, the Czech Republic and Denmark and had frequent contacts, support and substantial funding from internationally financed projects. This improved their laboratory equipment, as well as helping finance their activities.

In sum, one striking feature of Reflex is the remarkable continuity in its long term activity. Half of its "career" was spent at a time when very few environmental groups operated in Hungary, and it maintained their activities, virtually unchanged, in the new political circumstances. The other important point is that Reflex put the problem of global environmentalism before any concrete issues when it was established: "at the time we felt we should concentrate on long term plans. Environmental changes take a long time to bring about" (Gábor Szűcs, p.2), and concentrated their efforts on educating people, especially the younger generation. It was only at a later stage when they embarked on concrete actions.

And, finally, Reflex was chosen here in order to illustrate green activities outside the dominating capital. This proves that social movements are not confined to the centre. More than this, Reflex actually seemed to develop a particularly international profile in two respects. Its concrete contacts with Danish, Czech and Austrian colleagues turned out to be very fruitful for the movement. And secondly, Reflex actually shows a more "advanced", "German-like" approach to environmentalism. They seemed to be more informed and influenced by the more abstract environmentalist ideas, which developed mainly in German speaking territories, than any other movement. They did not come together as result of a concrete urge but on an unusually wide, long term and abstract basis, which is not typical in the Hungarian context.

Case Studies: Conclusion

I have described four different environmental movements, all of which existed in Hungary in the early 1990s. Some had a history going back to the 1980s (Danube Circle, Reflex), others came into existence with the "tide" of rapid political changes during the regime-change (Air Group) and became the best known in Hungary at a time when others, such as the Danube Circle, were already on the decline. Some of these movements were local (Green Future, Reflex), representing a particular district or region of the country, while others were national (Danube Circle, Air Group).

There were a number of environmental movements, apart from the ones I have chosen to describe in detail, which would be equally interesting to analyse and perhaps should be at least mentioned at this stage. Among them was a local environmental movement, in a village called Ófalu, in south-western Hungary, which successfully fought against a national government plan to build a nuclear waste storage site just outside the tiny village, which has been documented by Juhász, Vári and Tölgyesi (1993). Another local movement, in Buda, called the Fadrusz street movement, is still fighting. It wants to stop the government's plan to build a new bridge in the South of Budapest which would channel more heavy traffic into the area. Thirdly, I did not choose to analyse the only social movement in Hungary which is attached to a political party. This is the movement of Socialist Greens. It is so peculiar in its attachment to a political party that it is atypical of the overwhelming majority of social movements which specifically avoid being associated with political parties. Hence it was not felt appropriate to chose it. The information, however, deriving from our fairly detailed knowledge of these movements, as well as those detailed in the main part of this chapter, will be used in the next chapter which will analyse the major characteristics of Hungarian environmental movements. The environmental movements I presented therefore were carefully chosen to represent different types of movements which were typical for Hungary in the early 1990s. Here I have only introduced them by describing the different characteristics of the movements. I now turn to their analysis.

An Analysis of Environmental Movements in Hungary

The emergence of the current environmental movements can be traced back to the "euphoric" period of 1988 except for the pioneering exceptions which came into existence a few years earlier (Danube Circle, Reflex).

The Participants

The participants in the environmental movements, whether they were local or national, came from a particular group of society. They are mostly educated people, often with degrees in natural science. The core of educated people is often joined by housewives and retired people, and many environmental movements pay special attention to the younger generation, including those of school age. The age range therefore runs from students, or even secondary school pupils to the retired, but the most active members are often middle aged. The proportion of women in environmental movements is very high: in fact they often constitute the majority of participants.

When asked specifically, most activists described their movements as "mixed, containing a wide variety of people" from the unskilled worker to the retired manager. At the same time, however, well educated people are generally more valued by the movements as they are looked upon as potential experts who can contribute to coping with specialized matters needing legal or scientific expertise for example.

Those who joined environmental movements right at the start could have two reasons for seeking participation: political or environmental. A number of activists attached themselves to environmental movements as forums of political activity but as soon as political parties became legal and active they left social movements seeking political careers in political parties and sometimes subsequently in the government or in the civil service. "We, as a social movement, were acting politically speaking as catalysts. Many people joined us at the beginning because it was a political action, a form of opposition," argued an activist (Katalin Bihari, p.6). The majority of movement participants, on the other hand, saw themselves as environmentalists whose role was to support a non-party, non-governmental organization: "We, those people who are involved in environmental movements, believe that social movements belong to the domain of civil society and should not be confused with party politics. They are two different things" (Margit Varju, p.4.). This type of participant stayed on in the movements as loyal activists.

Movement participants consist of three "circles". Firstly, there is the core of the most active members, typically around dozen people, who devote most of their time and energy outside work to the movement. If the movement's financial situation allows, some of these core members become full or part-time paid staff members.

The second "circle" is the group of activists who cooperate on a regular basis and are often registered members of the movement. They often pay a symbolic membership fee as well. This circle can number from around fifty up to 300 people. The third "circle" is a larger group, often numbering a few thousand people, who are ready to participate in demonstrations to express their support for the movement. Beyond these circles are the sympathizers. The movements usually find it difficult to estimate the numbers of sympathizers. It is only at election periods or other major events when it becomes clear how many people sympathize with the movement among the otherwise more passive section of the population. This number could run up to tens of thousands of people.

The Movements' Leaders

The movement leaders are a frequently discussed question in the theoretical literature. The *resource mobilization* theory argues that they are mostly well educated and higher positioned individuals, with a strong motivation towards upward social mobility, which many seemed to achieve by simply becoming movement leaders (Oberschall,1973; Zald and Garner, 1987). The *collective behavior* school emphasised that socio-economic position on its own is not enough to lead a group of people; personal attributes of leaders, such as personal charisma are necessary too. Leaders who were well accepted, liked and respected for their expertise maintained the movement successfully (Smelser, 1962; Killian and Turner, 1972). Both schools assume a hierarchical type of leadership with one particular person as the main leader. *New social movement* theorists, on the other hand, emphasize group leadership with carefully divided responsibilities which are discussed and decided by the collective leadership on a regular basis. The tasks are delegated to the right person in the light of the different skills and personal characteristics of the different members within the collective leadership. In this kind of leadership it is the duty of the entire leadership to maintain a good relationship with the wider circle of activists to achieve solidarity. New social movement writers also emphasize the special role of middle class, well educated people in new social movements, such as environmental ones (Offe, 1985; Brand, 1990).

In the Hungarian case, we see examples supporting all three arguments in different situations. The leaders of movements are overwhelmingly well educated people in every case. Even if the movement itself attracts a fairly wide range of people in socio-economic terms, the leaders themselves are selected from those with the most expertise in some subject. There was no exception to this tendency. The leaders also see their position as an achievement in social mobility terms.

A wide spectrum of different age groups were represented among the activists but the leaders were usually middle aged. As mentioned above, women's participation in environmental movements is very high, but when it comes to leadership the different groups have shown different patterns. The Danube Circle, Air Group and Reflex had a male dominated leadership, but Green Future and other groups such as the Fadrusz Street movement and the Green Socialists had women leaders. In some Hungarian environmental movements women are, therefore, as well represented in the leadership as their proportion in the membership would suggest, but in at least half of the movements women are significantly under-represented. Einhorn (1993) and Voronina (1994) draw attention to some new trends in women's participation in the public sphere and political parties. They argue that it has decreased radically all over east-central Europe and that the level of female political representation has also fallen drastically. The trends in women's participation in environmental movements in Hungary both support and contradict this picture: women are active politically by participating in environmental movements in a majority, but they are not proportionally represented among the leaders in every environmental group.

Some Hungarian environmental movements had a hierarchical leadership structure and others a more collective type. Among the movements which had a hierarchical leadership structure the leaders' personal characteristics were looked upon as fundamental elements: charisma and expertise were the two most important ones (Danube Circle, Air Group). In movements where the leadership was collective (Green Future) *New social movement*-theorists' findings could be applied. The right person carried out the right task and movement solidarity was achieved by the entire leadership. The role of leaders was essential in any case in the survival of the movement. They had to be resourceful, full of initiatives, good organizers, respected persons, good negotiators vis-a-vis the authorities, and good at relations with movement members and with the public. Whether this was achieved on a hierarchical basis or collectively, did not make much difference to the movement from the point of view of its survival.

Social movements do not reward activists in materialistic terms. One of the rewards they can offer in return for many hours of voluntary activities is personal relationships, a certain feeling of "belonging". Social movement leaders had to be skilful in handling and managing people well by creating an atmosphere which was attractive enough to keep the participants together. Most surviving environmental movement leaders showed a remarkable talent in achieving this.

The Movements' Goals

Hungarian environmental movements mostly came into existence because of a particular, concrete goal they wanted to accomplish: to stop a major project on the Danube (Danube Circle), to clean or at least improve the quality of badly polluted air (Air Group), to divert traffic from a mainly residential area and to stop the construction of a bridge, which would attract further traffic (Fadrusz Street), to prevent a nuclear waste dump (Ófalu, Baranya), to close down a polluting factory and to divert a motorway route from a residential estate (Green Future). Even movements which originally had no concrete objectives developed one in the course of their actions such as pedestrianization in the town centre, traffic diversion, and protest against the Danube dam (Reflex). Whether these goals were achieved fully, partially or not at all, most movements came to the conclusion that their concrete objectives (which often were "not in my backyard" type claims) were fairly narrow and not political enough in a wider sense. Though movement members have diverse political views in terms of party political affiliation, there was consensus among most environmental movement activists that becoming more ecological in general terms, and simultaneously more political, was the right way ahead for the movement's development. Those movements which originally had only concrete goals at a later stage widened their horizons towards "green" thinking (Green Future, Fadrusz Street, Danube Circle) while other groups adopted a wider, green agenda from the start (Air Group, Reflex). As a result, all surviving environmental groups became more "global", pursuing a strong environmentalist agenda, even if it meant a change in the course of their development from a protest group with a concrete goal to an environmental movement with a wide range of projects. Conversely, those "not in my backyard" type protest groups which did not become interested in converting into environmental movements with a wide range of green interest died out, after they had achieved their concrete goal (Ófalu).

The question is whether surviving movements in eastern Europe followed the route of institutionalization and professionalization in the course of their changing character, as is often observed in western countries. In the case of environmental movements in Hungary there is no evidence of this. It is true that professionals are especially sought after in movements but any person with the right personal qualities was welcome. The fact that movements were on occasions funded from government funds did not necessarily lead to "institutionalization". It only allowed them to employ a few activists for moderate fees, based on short-term contracts and on a very insecure financial basis. No political strings were, however, attached to these government funds which were allocated by all-party parliamentary committees. Two other reasons why institutionalization did not take place were that movements maintained a strong principle of independence even from political groups, let alone bureaucratic organizations, and secondly, that the authorities did not attempt to incorporate them. (The question of the authorities' relations with the movements, including conflicts with them, will be discussed in chapter 7.)

Conflicts

Conflicts are essential parts of social movements as they always challenge something in the existing system. But conflicts can be internal as well as external. The two types of conflict can also relate to one another. The very origin of a movement is usually based on external conflicts which can pull movement participants together in the first place. They can add to the group's cohesion by strengthening it, but can also tear the movement apart. In the description of the different Hungarian environmental movements in the earlier part of this chapter I showed examples of both cases, sometimes within the same movement. In this part of the chapter I will concentrate on internal conflicts, and chapter 7 will deal with the movements' conflicts with outside bodies, such as authorities.

Instances of internal conflicts were given earlier in this chapter, for example, describing one of the local movements in the outskirts of Budapest, the Green Future group. Having achieved the closure of a huge chemical plant, Green Future first experienced a sort of euphoria which had a strong cohesive effect on the movement but did not prevent conflicts from developing among the activists little later:

> We were united with some kind of "fanaticism" when we heard about the closure of the chemical plant. But unfortunately later there were conflicts among us. Personal antipathy, which was concealed by the group's success, led to some internal conflicts. A financial windfall also caused conflicts and led to accusations among us about whether any of us had mishandled the money. (Ágnes Hársfalvi, p.12)

There were frequent rivalries and personal conflicts within the Danube Circle as well. Again, both success and failure contributed to them. Personal conflicts developed among the leaders after they were awarded the "Alternative Nobel Prize" and again when the movement was at its zenith of popularity. Similarly, the loss of hope of achieving the total abandonment of the construction of the dam provoked conflicts, too. Most of the movements, however, learned to deal with conflicts and absorbed them. In some cases this even united them (Ófalu). Personal conflicts were either avoided or kept at bay in the majority of the movements (Reflex, Air Group, Fadrusz street).

One major potential source of antipathy was political. At the time when most movements came into existence political parties were still in an embryonic state. The common feeling of opposition towards the regime was the most characteristic element both in the movements and the new political parties. But as the political parties developed so did movement members' affiliation with them, which became very diverse. This, however, did not lead to direct conflicts among the movements. All movements emphasized and practised great tolerance regarding members' political views, although it should perhaps be mentioned that no movements faced far right political views among their members as in Russia (see chapter 6). Political diversity did not constitute a problem in any of the Hungarian cases.

Success

When political parties think of success it is always related to their popularity in opinion polls and ultimately electoral success. As social movements do not aim to win elections they think of success in different terms which are more difficult to define or measure. Social movements aim to achieve goals, which could be short or long term, or both. The most tangible success of course is when movement manages to close a factory, stop a road being built, divert traffic, increase petrol tax, change a major project or stop a nuclear waste dump being opened near them. These have been achieved by many movements (Green Future, Reflex, Air Group, Danube Circle, Ófalu movement) in Hungary. However, social movements with broad general goals are more likely to survive, as argued by social movement theorists of the *resource mobilization* school (Zald and Garner, 1987), than those with specific goals, because a movement could vanish following success. Environmental movements in Hungary confirm this argument in two respects. Firstly, it is true that the only movements which survived for a longer period were those which broadened their goals and became environmental movements with a wide ranging green agenda. Those which achieved quick success but did not "develop" ceased existence soon after they achieved success (Ófalu movement). Secondly, this connection was recognised by the movements themselves. Having become aware of the potential constraints of having a too narrowly defined goal, as when movements were set up around a concrete grievance, many of the movements widened their interest to secure their own survival, in order not to become victims of their own short term successes (Green Future, Danube Circle). Some movements came into existence with both a broad and a concrete aim from the start and when they achieved success it was not perceived as a reason to disband but as one of the many aims to be achieved over a long term (Air Group, Reflex).

A recent paradigm within the social movement literature, the *cognitive approach*, has been applied specifically to environmental movements and sees success in terms of the movement's capacity to spread environmental consciousness (Eyerman and Jamison, 1991). The movements which transformed themselves into long term, more general green movements (which form the majority of movements in Hungary) viewed success in terms identified by the authors of the *cognitive approach*. Environmental awareness, educating people generally and young people especially, was or became a central focus of many environmental movements (Green Future, Reflex, Air, Danube Circle, Fadrusz street, Green Socialists). To change attitudes towards the environment as a whole was perceived by the movement activists as success. "The most important success in my eyes is the fact that people's perception has changed tremendously. People look at at nature, environment, environmental questions, waste, etc. differently" (Éva Utassy, p.15).

Another aspect of changing attitudes, as part of success, which was emphasized by Hungarian movement activists, was that "ordinary" citizens learnt to stand up for themselves and represent their own views. It was also an important achievement for the activists that these environmental movements established a strong reputation and are respected by major political parties, the government, local authorities and the public. No political party could ignore successful environmental movements even if environmentalism was not part of their own political agenda. Environmental movements became very successful in influencing public opinion. It became politically wiser to deal with them by consulting them and taking their ideas into account. Even if environmental movements do not participate in elections directly their presence is important in indirect terms. If a local environmental movement chooses to support a candidate within a constituency during local or national elections this has an effect on the outcome of the election. The MP elected in the district where Green Future operates was voted for on the basis of his strong environmentalist views, and Reflex sent a green representative to the county assembly in the 1994 local elections.

One of the most important successes environmental movements achieved was that they became "centres" which the local population could turn to if they came across *any* kind of environmental grievance. People in Hungary learned as early as the early 1990s that they could organize themselves and collectively pursue environmental issues, and challenge any large companies or the political establishment, which was in sharp contrast to the situation before. Only half a decade earlier, in the mid 1980s, as explained earlier, people saw the solution to environmental problems as either lying with the "almighty" state or to be solved individually (Kulcsár and Dobossy, 1988).

Social movements often see their role as to change legislation or the institutional process of decision making (Jenkins, 1985). Only popular and well supported social movements can achieve important legislative changes in any democratic society, as did the Air Group, or influence government decisions, as in the case of the Danube Circle. Local movements could also achieve success when taking on the government (Green Future, Ófalu movement).

Thus success can be seen in several different forms ranging from concrete, tangible achievements to more long term ones. Hungarian environmental activists considered long term success, such as educating the population to think "green" or to stand up for themselves collectively, often as more important than concrete victories, which could even lead to the end of the movement.

Media

It is clearly recognised by the movements that being reported on in the local and national media is beneficial. It helps them to achieve the fame and popularity they need to successfully put pressure on the authorities or companies. The media provides the widest communication channel for the movements. However, the media is influenced by its own interests which can harm as well as help the movement. Journalists appear when they wish to report about subject and present movements in the light of their own agenda which can be misleading. Also the media often sensationalises cases or portrays their scandalous aspects.

Nevertheless all environmental movements emphasised that having been reported on by the media is on balance more beneficial than remaining unknown to the public. Local movements can be especially grateful for achieving national fame via the national media which equips them with a lot more power than if they were only featured in local papers. Communication with their "constituency" is more useful, on the other hand, by appearing in local cable television as well as local papers, as was the case with Reflex in Gyor and Green Future in one district of Budapest. Publicity by advertising on public transport vehicles was the basis of the fame of Air Group. This was a new approach in two ways: by utilizing a business interest (the public transport company in Budapest) and by using publicity. These were well recognised by this environmental movement and achieved the desired effect.

Appearing in the media on a regular basis replaced the "old fashioned" method of demonstrations, which were so popular and powerful in 1989-1990. Social movements were finding it increasingly difficult to mobilise masses for demonstrations in order to put pressure on local or national governments. Since this "heroic" period, it is the media which can achieve powerful political effect. The media can of course be openly antagonistic to the environmental movement and support the views of the authorities. This happened in the case of Reflex, when the local paper supported the local council against the movement. This, however, was a fairly isolated case in Hungary. Generally speaking the Hungarian media is (a) interested in environmental matters and (b) supports environmental movements in their efforts. Most activists felt that they had a very good relationship with the journalists who regularly came to report about them and that they did not have to make much effort to be reported on by the national or local press, television or radio stations. In addition most of the movements wrote and distributed their own publications in order to gain publicity for their activities and aims and to encourage people to join or sympathize with them.

Conclusion

Environmental movements in Hungary developed at a fairly late stage. Considering the fairly liberal political situation in the country, which developed from 1968 and gradually led to major reforms in the economy as well as the tolerance of more "constructive" criticism than in any other socialist society, it is surprising that civil initiatives were so scarce. The reason lies in the prevailing ideology which did not allow the organization of social movements and diverted people's interests towards individual materialistic achievements. This unusual policy, as mentioned earlier, was deeply rooted in a fear of a repetition of the 1956 events. As a result, Hungary became the society in the eastern bloc with the highest level of private wealth and living standards but with a very low level of opposition movements. Thus the more liberal political atmosphere only translated into individualistic actions as far as the overwhelming majority of the population was concerned, and organized movements, including environmental ones, became widespread only in the late 1980s. Even though environmental awareness developed fairly early, from the late 1970s, due to Hungary's limited isolation from the central Europe, people expected solutions to environmental problems to be either organized from the "top", i.e. the state, or to be answered by individual action. The idea of non-governmental organizations as an option for expressing public awareness and pressing for solution was absent.

At the time when the one-party system started to dissolve and opposition parties appeared on the scene, first in an embryonic form, and the "round table" negotiations were initiated, social movements also appeared on the scene mushrooming in their thousands within a very short period (1988-1989). Many of those which came into existence then still exist. Originally most of them were organized around concrete goals. In order to survive, however, the short-term goals had to be widened. The broadened horizon of the movements led them to adopt wider aims than "not in my backyard" (NIMBY) objectives.

Consequently both educating the population to raise environmental consciousness and to strengthen the movement politically became important aims of the movements. Concrete goals were not abandoned but became part of the objectives. Once they were achieved, new goals were adopted. Environmental movements in Hungary also achieved political respect. The absence of successful green political parties also contributed to the movements' political strength. They have a direct or indirect influence on the major (non-green) political parties in the sense that where there is an environmental movement the local political representatives have to express a clear view on environmental issues and are rarely elected without supporting green issues. Most of the major political parties, however, do not consider the environment as their most important problem, not even the party (the Socialists), which itself has an affiliated green movement (Green Socialists) and has beens in government since 1994. Nevertheless, environmental movements do a valuable job in changing the Hungarian population's attitudes towards green matters by drawing attention to them and keeping them on the agenda, as well as achieving concrete aims. We now turn to the Russian environmental movements.

6

Environmental Movements in Russia

The previous chapter examined Hungarian environmental movements, firstly describing them and then evaluating them. This chapter will analyze Russian environmental movements in the same order. The chapter falls into four parts. Firstly, I shall look at the causes of environmental problems and whether there is a growing awareness concerning environmental issues in Russia. Then I will look at the origin of environmentalism in Russia, the different ecological concepts prior to the Soviet period and under socialism. The third part of the chapter will describe several existing movements which all came into existence in the late 1980s. Finally, I will analyse Russian environmental movements.

Industrialization and the Development of Environmental Ideas in the Soviet Union

Industrialization under state socialism was pursued on an even larger scale and for a longer period in the Soviet Union than in Hungary or any other eastern European socialist country. The basis of the ideology, which led to this accelerated industrialization, was similar to that in Hungary, and had its origin in the Soviet Union. Both the Marxist approach and the cold war constraints were primarily Soviet ideas. The Soviet Union, being a country of enormous physical size, had a much larger land area with forests and other uncultivated areas than other eastern European countries. This vast natural area, and the cultural value attached to it, led to a romantic attitude towards the natural environment, which was not the case in Hungary. A good example of this is Russian literary novels and poems which express great distress at the deforestation and the shrinking area of natural beauty in the Russian motherland.

Environmental ideas, which developed in the western industrialized world, did not influence the Soviet Union to the extent they clearly did Central-Eastern European societies, because the Soviet Union was a lot more isolated from the west than Hungary, for example. The upsurge of environmental concern which developed both in the developed world and in some socialist countries in the 1970s was virtually unknown in the Soviet Union.

The Soviet Union's political and economic situation was also significantly different from that in other state socialist societies. The reform "winds" of the short-lived Khrushchevian period were reversed under Brezhnev and living standards remained much lower than in countries like Hungary. The accumulation of private wealth was not encouraged at all, the second economy was not legalized, and ideology gave priority to communal, as opposed to individual, thinking. At the time when "subbotniki", (so-called voluntary community work) were long forgotten in Hungary, in Brezhnev's Soviet Union it was still widely pursued as a way of encouraging people to do more "optional" work for the community in their spare time. The official collectivism implemented through the strong grip of the local party apparatus did not, however, succeed in achieving its aim. By the first half of the 1980s there was evidence even in the Soviet Union suggesting that the ideology did not work. A study investigating work and leisure activities in 1984 concluded that the majority of the urban population were tired and apathetic, and felt they had exhausted their moral and physical resources (Abankina, 1986 quoted in Yanitsky, 1993b). It found that 75 per cent of the respondents were very pessimistic, and felt they could not hope for any changes in their lifetime. The respondents felt they had no strength to implement changes individually and there was also a decline in interest in work. Instead various forms of "escape" were on the increase (alcoholism, drug abuse, retreat into private life, migration from one part of the country to the other in search of higher wages or better opportunities). Faith in the values of socialist ideology was disappearing, leading to an ideological vacuum and psychological tension (Yanitsky, 1993b).

Environmental consciousness developed only very slowly, even though the Soviet Union experienced the biggest man-made environmental catastrophe at Chernobyl in 1986. A survey conducted in 1989 (Doktorov, Firsov and Safronov, 1993) found that low income, inflation, shortage of food and consumer goods, and housing problems worried people more than environmental issues, which came only fifth in the list of the most acute problems affecting Soviet people at the time. But environmental problems were given more importance than problems such as bureaucracy, corruption, low standards of medical service, ethnic conflicts and degradation of public morality. Even though environmental issues were not the most important concern for the Soviet population by any means, at least they were mentioned, according to Doktorov's 1989 survey, among the first five alarming social issues. When asked specifically about the state of the environment in their country the overwhelming majority (85 per cent) of respondents expressed concern about the ecological situation of the Soviet

Union (Doktorov, Firsov and Safronov, 1993). In other words the overwhelming majority of the population had some knowledge and concern about the state of the environment by the late 1980s, but they did not regard environmental problems as the most important problems in their everyday life given that the state of the economy.

Environmentalism in Russia

Environmentalism in Russia goes back to the pre-Soviet period. By the time of the revolution in 1917 a modest conservation movement was already established. The need for nature protection was recognized by journals, societies, a quasi-governmental commission and an informal network of professionals. Weiner (1988) identified three basic positions among conservationists: *pastoralist, ecological* and *utilitarian*.

The basis of the *pastoralist* approach was strongly antimodernist: it was repelled by modern industrializm and sought to return to an idyllic rustic "golden age" in "mother-Russia". Humans were viewed as "children of mother-nature' and it was argued that "industrialized human beings" had become de-natured. The pastoralist view saw contemporary humankind as a "pathological element" which disrupted the pre-existing "harmony between nature and people". The pastoralist views, which developed in Russia in the late 19th century and were still influential at the turn of the century, were strongly influenced by contemporary German neo-romanticism. The strong German patriotic/chauvinist accent was also influential on the Russian pastoralists.

The second early position was the *ecological view*, which was strongly materialistic in contrast to the pastoralist one. Followers of the ecological view were also deeply worried about the consequences of civilization, the breakdown of the natural eco-system, but on an anthropo-centric basis: worrying about the future of human kind in a destroyed natural environment. They argued for strong policies in economic matters and resource use. The group consisted of natural scientists almost exclusively. In the early Bolshevik period enlightened leaders, such as Lunacharsky, supported the ecologists, but later, under Stalinism, the state-technocrats turned against them. The early ecologists were politically progressive in their views. Rooted in their philosophy was a strong element of protest against the emerging new order in which, in the name of collectivization, intellectual autonomy could be lost. Nevertheless the early Bolshevik period encouraged and accommodated these groups because they were enlightened, materialistic and scientific. Lunacharsky, who was committed to humanistic education, cultural pluralism and intellectual autonomy and promoted values on this basis, warmly welcomed the ecologists who sought to provide a scientific explanation for complex natural phenomena. Stalin, however, turned against them.

The third position, *utilitarianism,* was also rooted in the pre-Soviet period, but was especially favoured under Stalin. Utilitarians defined resources narrowly, based on the limiting criteria of economic utility. They favoured the idea of growth and embraced technocratic priorities, excluding recreational and aesthetic amenities and all things (living or non-living) which had no economic value. Utilitarianism triumphed from the late 1920s, when Stalinism became the ruling power (Weiner, 1988).

At the time when civil initiatives went through hard times, during the Stalinist period, ecologists were also rejected. It was the Khrushchev era which allowed civil initiatives to surface again, among them environmental activities as well. The Nature Protection Squad of the Moscow State University, founded in 1960, was a pioneering organization. It started as an organization of students and lecturers of the Faculty of Biology but grew to become a national movement surviving difficult periods after the Khrushchev "thaw". The Squad trained brigades of people who then became active all over the country (Perepjolkin, 1997; Yanitsky, 1993a).

In more recent times it was the final years of the Brezhnev period, when the symptoms of the decay of Soviet Communism reached the point when activities outside the control of state increased. Sharp criticism within the party, as well as outside, was voiced from the mid 1970s, including environmentalist criticism, but only in the period of perestroika did civil initiatives become popular. The series of ecological disasters which became public for the first time, among them the Chernobyl catastrophe, speeded up the growth of interest in ecological safety, but only to a limited extent. It was not the large-scale catastrophe of Chernobyl which provided the change in people's thinking in terms of political mobilization, but the political environment created by Gorbachev's perestroika and glasnost policy. This is why from the late 1980s so many ecological movements were organized, as were other civil initiatives at the time, in the rapidly reforming Soviet society. Some environmental groups which came into existence at the time in the Soviet Union had strong concerns about local problems while others pursued national objectives.

Environmental Movements Prior to the Perestroika Period

Environmental movements prior to the perestroika period, which started in the mid 1980s, existed in the Soviet Union in the form of local initiatives, neighbourhood groups, and clubs, such as university clubs. Some were concerned with great natural features, such as Lake Baikal, as discussed earlier, the Siberian rivers or the Aral sea. Voluntary patrols for protecting the environment had been active for decades, since the period of

Khrushchev. Environmental catastrophe, such as Chernobyl, however, did not give birth to organized anti-nuclear protest (Yanitsky, 1993a), even if people's opinion did turn against nuclear power stations, as testified by opinion polls (Doktorov, Firsov and Safronov, 1993). According to Doktorov's survey, conducted in August 1990, only 13 per cent of the Soviet population supported the use of nuclear power while 54 per cent rejected it. But environmental protection was more a continuation of the romantic cultural notion of "preserving the once unspoiled Russian mother-land". Nevertheless, the ecological groups which were around prior to the perestroika period provided experience in organizing environmentally related group activities which was to be useful later, but this was as much as could be achieved given the political circumstances of the Brezhnev period.

Political repression did not allow very much scope for oppositional civil initiatives under Brezhnev although the regime could never fully succeed in silencing it. The case of Lake Baikal illustrates this well. Soviet-type opposition, in a regime which did not allow civil initiatives to operate freely, turned to different methods. Firstly, experts, such as biologists, hydrologists and geographers drew attention to the alarming prospects for the lake and its environment when the plan for the paper-mill construction emerged. Secondly, well known writers and academicians joined the growing opposition. Finally, local people organized protest groups, demonstrations and successful petitioning. The latter, however, did not occur in the case of Lake Baikal prior to the Gorbachev period (Wilson, 1993).

One result of Brezhnev's policies, by forcing an ideology down the throat of an unwilling population, was that individualism and political apathy became a political escape route for Russians as it did for Hungarians. When, however, a new style of policy, Gorbachev's glasnost, was introduced, which undoubtedly woke and shook the Soviet Union to an unprecedented extent (Sakwa, 1990, p.1-9), political apathy vanished.

Gorbachev's policy was most successful in highlighting and bringing into the open certain facts which were well known to all within the Soviet bloc from personal experience: the enormous degree of corruption within the state and party bureaucracy, the lack of good professional conduct, the inadequate coordination of the economy to increase production and reduce waste, and improve services, and so on. All these existed in the country in contradiction with the official Brezhnevian propaganda and were admitted by Gorbachev for the first time. As a result, political apathy was suddenly transformed into a political upheaval with (according to some calculations in 1988) around 60,000 civil initiatives (Yanitsky, 1993a), appearing on the scene simultaneously.

The Appearance of Environmental Movements

The upsurge included numerous environmental movements. They were tolerated under Gorbachev, but not legalized for a few years, just as in Hungary in the last period prior to 1989, but that did not prevent them from mushrooming all over the country. The environmental movements of the "transitional period" of 1985-1991, that is during Gorbachev's perestroika, played a very important political role. Unlike in Hungary, political parties in the Soviet Union did not develop parallel to civil initiatives due to Gorbachev's reluctance to give up the primacy of the Soviet Communist Party till the very end of his presidency. This hindered the development of political parties and the introduction of a radically new political regime in the Soviet Union. The Gorbachev "revolution" therefore managed to shake up the regime and introduce steps towards democratization but did not let it develop to the full. Within the Soviet Union, Gorbachev was both radical and too conservative in political terms. His innovative efforts aimed at radical democratization but turned into damaging conservativism by keeping the Communist party's exclusive role and preventing Russia from becoming a multi-party system. This unnecessary delay caused a limbo situation within the economy as well, leading to a substantial decline in living standards and an economic despair in Russia. Gorbachev's political ambiguity led to general discontent and provoked the coup of August 1991 but his innovative ideas in pursuit of democratization were very important.

Environmental movements under Gorbachev became a focus of "safe" opposition. A similar phenomenon occurred in Hungary at the very beginning of the transitional period, in the late 1980s, until people realised that there was no danger in openly participating in anti-communist parties or groups. The feeling of confidence that there could not be political retaliation occurred at a much later stage in the Russian case because of the harsher political atmosphere there in the past, and environmental movements remained the vehicles of general oppositional forces for a lengthier period, until well after August 1991.

Environmental Movements in Russia: Case Studies

In this part of the chapter I will describe representative examples of Russian environmental movements. This will include the Socio-Ecological Union and the Moscow Ecological Federation, which are both umbrella organizations, representing a number of local environmental groups, both in Russia and in Moscow. I will then introduce the Russian "branch" of Greenpeace International, which was set up with their help and is called Greenpeace Russia. Local environmental movements will be represented by two movements. The first is called Bitsa, a neighbourhood protest group, a local movement against the Northern thermal electric station, and the other is called Ecopolis.

Among Russian social movements there is a strong tendency generally, not only in the case of environmental ones, to form umbrella organizations which, the movements feel, can be more effective in dealing with higher levels of bureaucracy. First I look at one such federation.

The Socio-Ecological Union

The Socio-Ecological Union (SEU) is a national organization with members all over Russia. It also has active contacts with members of environmental movements in the former Soviet Union who were previously members of SEU.

The Origin of the Movement

The origin of the movement goes back to the already mentioned Nature Protection Squad which came into existence in 1960, and which was very influential on the SEU, both ideologically and in its organizational principles. Several members of the Union started off in the Squad. The contemporary movement, on the other hand, is different from its predecessor in its philosophy. It is an "anthropo-centred movement" (Weiner, 1988) which means it has a strong concern with the protection of the health and living conditions of people, unlike the Squad, which was a more romantically inclined nature preservation organization fighting against poachers and other individuals harming the Russian countryside.

The Union sprang up after 1986-87, when protests against a major project to divert the flow of northern Siberian rivers south, to provide water for the irrigation of the cotton fields in Central Asia, became strong and finally successful. Many of the present movement members took part in environmental protest against the government for the first time during this action.

But there were three other important phenomena contributing to the foundation of the movement. One was the increased availability of information on environmental matters. Glasnost gave more opportunity for the press to report about protest actions and a degree of possible success. That encouraged existing environmentalists and mobilized new ones. The other was the Chernobyl disaster and other man-made catastrophes which were reported for the first time. These made it clear to many people in Russia, far beyond the existing small circle of environmentally inclined people, that life threatening events were common in Russia and that something had to be done about this. Thirdly, the fact that Russia, which before perestroika was fairly closed off from contacts with western societies, opened up and international contacts started to be built on a fairly rapid scale. This led, in the case of the Socio-Ecological Union, to "exchanging notes" and developing long term relationships.

The Participants

The participants in the SEU were in part former activists of the Nature Protection Squads who had been students in major cities like Moscow, Gorky and Novosibirsk. The Union maintained the organizational principles of the Squad in the sense that it brought together groups of movement members in different points of Russia, mainly in urban areas. In 1989 they had 1,000 member groups. The most active participants were reported to number 300 but an action would attract 3,000 people. One of the co-chairmen of SEU claimed that overall "counting signatures, participants in meetings, discussions and demonstrations there were around a million people involved in the movement." This could be a slightly exaggerated claim, even if it is true that at the height of the period when demonstrations were the most important forums of protests, a sizeable mass of people was active in environmental demonstrations on a regular basis. Later all movements changed their main forms of actions.

The Leadership

Russian social movements often develop a strong resentment towards hierarchical structures which are so typical of the Russian and Soviet bureaucracy. These feelings were very strong in the SEU and led to the decision to develop a horizontal rather than a vertical structure within the organization. SEU built up a network of horizontal links with all the groups involved in it emphasising that each member group has equal rights. This somewhat romantic egalitarian principle is adopted by many movements as a response to past experiences. The leading role is in the hands of several coordinators, who are in charge of various programs and activities as well as information within the movement. They have tried to set up a council of elected representatives, but it has only a symbolic function, and in practice does not operate. Recently there has been an increase in the numbers of paid staff funded by foreign environmental movements. Most of these staff work in the Moscow centre of SEU. There is no membership fee or formal registration for the individual members in order to avoid being seen to be too bureaucratic. SEU is, however, legally registered with the authorities but this is considered by the activists to be a pure formality.

The Goals of the Movement

The goals of SEU are diverse, due to the fact that it is a national organization with many local groups which individually formulate their own concrete aims. One of them is to make sure existing environmental legislation is observed in Russia, a role resembling the one the Squad pursued during its activities, though the Squad only ever blamed individuals, whereas SEU sees environmental problems as structural ones. Now they talk about "unbearable conditions" as a result of industrialization, collectivization, nuclear-power stations. "Previously we only talked about saving nature, now the talk is about saving human lives" (Cherkasova, an SEU activist, p.2). SEU does not want to maintain close relationships with any political party; they feel that ecology is apolitical in party political terms, though recognizing that everything they do is political in a wider sense.

Success

Success is mainly perceived in abstract terms by SEU. "The main success of the movement is that it is a fully developed part of civil society, its active institution is in the making and it has been recognized by the authorities. The authorities show to the movement respect and consider our opinions." (Zabelin, activist, p.9). Another achievement in the judgement of the SEU is that they became "part of the international arena".

Media

The media played a very important role in the formation of the SEU both by providing information about other movements and by reporting about their activities. In 1987-89 the national press, as well as the major television news program "Vremia", reported about their involvement in the protest, connected with a power station on the River Katun, in Siberia. Lately the foreign press has shown more interest in their activities than has the national press. But this is not a "personal" conflict between the Socio-Ecological Union and the Russian press. It is a general tendency in the Russian media, which shows very little interest in environmental problems generally. On the other hand, SEU activists are very active in publishing their own local newspapers all over Russia—there are thirty to forty papers edited by their member organizations.

To summarize, the Socio-Ecological Union is special in the sense that it is an umbrella organization, spreading all over Russia with many member groups located in different, mainly urban areas. It is built on the experience of a conservationist society with a history going back to the 1960s. It has accommodated not only previous members of the Squad but its network-form as well. It has become well known beyond the boundaries of Russia, which contributes greatly to its financial success. The activities of the Union are wide-ranging. At the centre of the SEU they tend to concentrate on abstract goals, such as acting as strong representatives of their member groups vis-a-vis the national authorities, and coordinating information flow among the member groups. The concept of success is also abstract in the SEU activists' definition: concrete success, they argue, should be achieved at regional and district levels by their local groups.

The other umbrella organization representing local movements in the Moscow area is the Moscow Ecological Federation.

The Moscow Ecological Federation

The Origin of the Movement

The origin of the Moscow Ecological federation (MEF) goes back to the late 1980s when many local environmental groups were mushrooming in Moscow. Some of those who acted as contact persons among these local movements felt that they would become more effective if they were united in a Moscow-level federation. It was also felt that there were ecological problems at the city level which could not be solved by neighbourhood groups. In the winter of 1987-88 many of the leaders of local groups met and by the second half of 1988 they came to the conclusion that they should set up a federation. This was finally founded in April 1989. The Moscow Ecological Federation was officially registered as part of the All-Russian Nature Protection Society, which was a pure formality, but it provided the activists with meeting places.

The Leadership

The leadership is practically one person, though theoretically there are three co-chairs. The only active leader is a woman who worked as an electronic engineer for more than twenty years. Then she joined a neighbourhood environmental action-group and later became elected as co-chair of MEF. She was also involved in the all-union Green Movement for two years and after that she became the deputy head of the of the Centre for Coordination and Information of the Socio-Ecological Union, where she works parallel to her MEF activities, which is an unpaid position.

MEF is a confederation of local movements throughout Moscow with a coordinating board which consists of representatives from each member movement. They hold annual conferences, with around a hundred member representatives present at a time, and coordinating board meetings with thirty representatives, on average once a week, or occasionally less often, depending on how many problems there are to discuss. Several dozen experts help them in their work and there are sixty activists associated with the centre itself. They claim that through the local group members they have 7,000 or 10,000 participants. Lubov Rubenchik, the leader, complained, however:

> I am not happy with the size of the Board because in reality I am the only person who has to do all the coordinating work. We have two co-chairs beside me, but they don't do any coordinating work. One of them became a city councillor, the other one is no longer involved [in the movement]. (p. 5).

Most of the activists are middle-aged or older, and most are well educated. Their political affiliations are very diverse. As far as the recruitment "policy" is concerned, it is believed by the activists that a movement needs experts rather than masses:

> Mass enthusiasm can discredit the movement if it is not based on scientific expertise. Experts can be employed by governmental institutions or independent bodies. Each demands full loyalty. Therefore independent experts are needed to tackle governmental claims, people who are experts in their fields and are active citizens in grassroots, non-governmental organizations are of most use for movements (Rubenchik, p.3).

The Goals of the Movement

The goals of MEF are threefold. Firstly, they aim to take up all Moscow-related ecological problems. The Federation wants to secure the right to participate in decision-making at the city level both in concrete plans and over long-term planning. They also wish to ensure that the authorities act in line with existing environmental legislation which, they argue, is often ignored. Secondly, MEF gives all the support it can by backing local movements in their struggles. Thirdly, it tries to coordinate the activities of local movements so that a flow of information exists among them.

On a more concrete level, the MEF was very active in opposing any new houses being built in Moscow. They argue that Moscow is overpopulated as it is and does not need any new houses, and that enterprises should be moved out of the capital to attract the population to other parts of the country. At present construction of new houses and enterprises is attracting yet more Russians into the capital. The city cannot cope with any more migrants or even with its existing population, according to MEF activists.

MEF is also trying to set up a database and information centre, which would allow them to advise firms how to change to environmentally friendly technology to avoid being fined. They hope that this part of their activities can grow into a commercial one, advising companies and charging them for their advice in order to generate some income.

Success

It is difficult to achieve success in a movement which identifies its goals in such wide terms. The MEF does not, therefore, claim any concrete successes: "when the government stops violating environmental laws that will be success. When courts will investigate cases, that will be success. When industrial plants will use filters, that will be success." (Vorobiev, p.9.) They consider it an achievement that they could participate in the evaluation of general plan for the development of Moscow. They are still struggling to influence the city council in its final decisions on concrete plans.

The Media

Media interest in MEF is sporadic. National newspapers seek to interview them only occasionally but MEF activists often feel they are misinterpreted. Green newspapers, however, frequently seek their help, by asking MEF members to contribute articles and to distribute green magazines, because they have difficulties in selling copies. One of the council members gives regular talks on Radio Russia on environmental issues. This sporadic publicity in the national media, however, does not contribute to strengthening MEF's position among the public outside movement circles.

To sum up, the Moscow Ecological Federation is, again, an environmental organization uniting local efforts and representing them at the city level. But since MEF has inadequate financial resources, which come mainly from western sources, they cannot afford to employ any paid staff. This leaves the bulk of the coordination to enthusiastic volunteers, which in the MEF case means to one person, and she finds it very difficult to cope. Thus MEF is not very strong at fulfilling its two major aims, coordinating local movements and representing them at the city level. Nevertheless MEF activists try very hard to provide as much support for local environmental movements in Moscow as they can and to influence the city's long term plans.

Now we turn to the third Russian environmental movement which aspires to be a national umbrella organization, but of a different kind. It is a "local branch" of a well known international organization, Greenpeace International, which covers most western European countries, and now has a group in Russia.

Greenpeace of Russia

Greenpeace of Russia is not the only organization which has international contacts. As we saw above, the Socio-Ecological Union is also helped by foreign funds. The fame, the skills and the organizational principles of Greenpeace International, however, provide an interesting case for Greenpeace Russia.

The Origin of the Movement

This is a fairly new movement. It was not organized as a spontaneous collective action in the late 1980s, like most Russian environmental movements, and it is not a grassroots movement by any means. It is a highly institutionalized operation, initiated by Greenpeace International, which wanted to expand into the former socialist world, and chose Russia as a target because of its international importance. The office was opened in 1990. Greenpeace International provided the equipment for the Russian Centre, but surprisingly their main source of income is not Greenpeace International but local members. Greenpeace International, however, set up an office with a director and several campaign managers. These people are in charge of different subject areas, such as nuclear power, disarmament, forests, toxic waste and the state of the ocean with special attention given to the consequences of fishing, closely following the pattern of the "mother" organization.

Not only they are not financed from abroad but Greenpeace International even targeted Russian people with publications and other products, like records, even before the office was set up in Moscow, and attached to them leaflets encouraging Russians to contact Greenpeace International and send substantial donations (by Russian standards) in American dollars.[1]

Participants

Greenpeace Russia has ten paid staff members, eight of whom are involved in the campaign work. Apart from that around twenty volunteers contribute to their labour to the activities of the Moscow office and 10,000 to 15,000 members contribute regular membership fees. Most of the paid staff are in their twenties and thirties and are well educated with university degrees in geography, geology, or sociology. Some of them are PhD students. As work is flexible, in the sense that they often work ten to twelve hours a day, and most weekends as well, it is an advantage to be young and single: "we have young people because we have got a lot of hard work to do, for married people it would be more difficult working here" (Sokolov, a campaign coordinator, p.8).

The Leadership

The leader of Greenpeace Russia is the director, who is Russian by origin, although he left Russia at the age of twelve, when the family was forced to leave the country. He is not a Russian citizen any more, but was sent to Russia because of his personal roots combined with his experience in western branches of Greenpeace. Part of his job is to pass on his skills to the local activists. He is highly respected by the staff for his knowledge and experience in the field. There is good cooperation among the different staff members as well, although their work is clearly separated by subject areas. As a result of this "parachuting" method, Greenpeace has achieved a Russian environmental movement which in organizational skills is far superior to any other movement in Russia. However, it is not based on organic development, which raises the question of the extent to which it is an institutionalized organization and whether it is a "collective" action in the same way as the other movements we studied.

The Goals of the Movement

The goals of the movement derive from two factors. The first is the fact that it was initiated by Greenpeace International, which had already developed certain areas of activity and a style of action. Greenpeace Russia therefore became a highly organized movement right from the start with well defined goals and well equipped offices in strong contrast to all other environmental movements in Russia. This, however, also meant that Greenpeace Russia became fairly institutionalized from the beginning.

The second factor derives from the Russian context. Greenpeace Russia campaigns against nuclear power and for disarmament, as in any other country, and to protect forests in regions like Karelia, the Northwest, and Central Siberia (Irkutsk and Krasnoyarsk region) where the timber trade is concentrated, often involving joint companies (Finnish, Korean) with excessive tree felling. It also campaigns against trade of toxic substances and the export of acid waste to poorer countries, like Russia. It tries to prevent environmentally harmful technologies from being imported to the country. The fishing campaign concentrates on the problem of overfishing of Russian waters, often by joint ventures. Fishing companies in which 31 per cent of the shares are owned by foreign companies can sell fish abroad without a licence, which they do in desperation for hard currency.

Greenpeace Russia concentrates its efforts on large issues which also attract attention in many other countries where Greenpeace International is involved. But it concentrates on their Russian aspects and especially on international projects and joint ventures. As one of the campaigners put it:

> Greenpeace is an international organization. That is a specific thing about it, for example my campaign against the trade in toxic substances tries to prevent environmentally harmful technologies and acid waste from entering the country. The nuclear power campaign also deals with international problems. Fishing and ocean problems are international too. The forest campaign keeps watch on the problems of foreign capital participating in the production and trade of timber. That's specific to Greenpeace. (Strigulian, Greenpeace Russia activist, p.2.)

Success

Success was achieved in a campaign organized against the South-Korean company Heido, which was involved in logging in the Far East of Russia, and violated several laws but escaped prosecution. Greenpeace Russia launched a campaign against the firm which roused the public and as a result the firm lost its licence. In other cases Greenpeace Russia draws attention to environmentally dangerous activities, like dumping nuclear waste in the Baltic Sea, or to the overfishing of the Okhotsk Sea and the cheap selling off of fish which they also view as achievements. They also lobby very hard to influence legislation.

Greenpeace campaigners also emphasized their role in changing the general attitude of people in Russia towards the environment. Given that economic problems are more important for people than ecological ones at the moment, changing people's consciousness and making them understand the ecological threat would be their biggest success, they felt.

The media

The media are involved in publicizing their activities; however, not the Russian media. Gathering information and passing it to the press occupies 50 per cent of his time, claimed one campaigner, but the Russian press does not pay much attention to their activities. It is often the more specialized magazines which use the information they provide, rather than popular national papers. On the other hand, the international media is in contact with them all the time. Apart from this Greenpeace International itself finances special publicity projects, such as a film on Russian forests: "The image of Greenpeace will attract the audience, the 'trade label' will attract the public's attention to the film" (Tsyplenkov, activist, p.12). This seems to work as a trade off, in which both Greenpeace International and its Russian branch, benefit.

Greenpeace Russia thus obviously differs from other environmental movements primarily in that it has never been a Russian grassroots organization growing to achieve international fame, but was set up as a highly institutionalized, and by local standards, well equipped, branch of a successful international organization. Despite, or perhaps because of that, it is popular in Russia, with a large number of supporters and is fairly successful considering how short its history is. Its survival does not depend entirely on local efforts but the full-time paid participants, as well as the group of unpaid helpers, show great enthusiasm which could help give it a long-term future. The young people working for Greenpeace Russia find their work extremely rewarding and the number of sympathizers is surprisingly large.

Local Environmental Movements

Apart from the movements which are concerned with national or all-Moscow interests there are many grassroots initiatives which operate within one locality.

The Movement for the Protection of Bitsa Forest

In contrast to the movements described above, the Bitsa movement is an example of a local protest with a very concrete goal. It has, however, become well known all over Moscow.

The Origin of the Movement

The origin of the movement goes back to 1987, when many environmental movements were founded throughout the Soviet Union. The movement emerged as a reaction against a concrete project, the plans to construct a zoo in Bitsa forest in Moscow. This forest is next to a modern housing estate and is the only green recreational area in proximity. The forest, argued the activists, is important in absorbing some of Moscow's heavy smog. Even if the movement became known all over Moscow, it remained a local movement.

Participants

The movement participants are people living in the neighbouring houses who are very attached to the forest. A lot of houses in the neighbourhood were built by enterprises where the working conditions were bad and employees developed occupational diseases. Most of them are older people, often retired and with a low income, who do not have any chance of a dacha or seaside holidays. The forest is the only leisure opportunity for most of the participants. This explains why most of the movement participants are older people. Young people under the age of 30 are extremely rare among them. The activists did not know each other previously but became friends as time went on. Some of them are lonely people, for whom movement activities are a way of socializing as well as feeling good about doing something for the community. Some of the movement participants are professional ecologists, but only a few.

The movement accommodates people with diverse political views. Some of them are strongly opposed to the new regime and maintain their Communist beliefs, while others are closer to political views represented by democratic Russia. This, however, does not create conflict among them. They do not have any contacts with political parties though some of the existing or former activists stood for elections on different levels from local councillor to MP.

The Leadership

The leadership is formal and well structured with the tasks well-defined and shared. The Bitsa movement does not share the reservations of the Socio-Ecological Union and the Moscow Ecological Federation towards hierarchical leadership. There is a leader and two deputies who are all described as charismatic persons and good organizers. They are also very good at communicating with the authorities which in the members' view is a crucial leadership skill. Major decisions are taken collectively by the so-called governing body, the core of activists of about fifteen people. None of them are paid for their activities or even reimbursed for their expenses. They had 800 active members when the fight against the zoo was on. Now the activity is reduced to the core activists on a regular basis but they claim that the movement can still count on the support of some five thousand people, more or less the whole neighbourhood. This was the number of people who attended rallies.

Conflicts

Recent conflicts, however, have divided the population in their support. Many of the locals wish to use the forest for building (illegal) garages for their cars. As the crime rate has grown considerably in Russia, and shortages of car parts have increased, one of the ways of replacing a part is by stealing it from a parked car. The thieves take every possible part from cars standing in the streets and sell them to desperate car owners. This situation has increased the demand for garages to keep private cars safer. This, in turn, divides the population between car owners, and non car owners who do not wish to sacrifice every plot of land in the city to build garages, for example in the forest or on river banks. In this neighbourhood approximately half of the population is "pro-garage" and the other half wants a "garage free" forest.

Another source of conflict was the local authority. It was originally hostile and antagonistic towards the movement. But in the late 1980s, in the last period of perestroika, when the "democratic atmosphere" effected even local authority bureaucrats somewhat, the movement's growing popularity persuaded them to change their attitudes. They started by accepting the activists' arguments, and later even supported the movement by providing some funding. This limited financial support was the only support Bitsa achieved during its whole existence. The movement has no contacts with organizations outside Russia, or even Moscow, but is a member of the Moscow Ecological Federation.

The Goals of the Movement

The goals of the movement are very clear and well defined: the protection of the Bitsa forest. Firstly, they struggled against the zoo. When it first became known that the authorities wished to construct a zoo in the forest, the consensus of local opinion was against it and the freshly organized movement gained a lot of support. After that the movement opposed plans to build houses and roads through the forest. At present their task is to make sure the decisions of the authorities concerning the forest are actually carried out. The latest fight is over the garage question and to maintaining the forest as a clean and green park.

Success

The Bitsa movement has achieved many of its concrete goals. The zoo project has been cancelled as have house building plans. In fact, the forest was given the official title of a "Natural Reserve Park" on 17 October 1991 by the city council, as a result of the movement-generated public pressure. This means that no construction is permitted in it. The only task remains to monitor whether the decision is carried out and to make sure that no construction without permission takes place in the forest.

Media

The national media only reported on the movement at its peak activity period. Since then, just as in other cases, there has been no reporting about their continuing activities, except for the local paper which runs articles about them on a regular basis, keeping the local population informed, but this does not spread knowledge about their activities outside the neighbourhood. The movement itself does not have financial resources for their own publications or even photocopying, as their only income over the years was a small fund from their local authority and personal donations from the public which in total was not very large.

In sum, the Bitsa movement is a good example of the numerous local movements which came into existence in the late 1980s. It is undoubtedly a fairly successful movement, which is rare among Russian social movements. It achieved all its objectives, which were clearly defined and supported by a large proportion of the local population. The Bitsa movement used the methods, most frequently used by Russian movements in the late 1980s, to put their views forward to the authorities in charge of the forest: namely mass demonstrations and rallies. Their popularity changed the local authorities' attitude towards them for a limited period around the peak of their popularity and from being opposed to them the local officials became the movement's funders for a short time. Nevertheless Bitsa remained a poorly resourced organization, relying mostly on participants' activities, contributions and enthusiasm. The movement continued to stay together even after it had achieved many of its primary targets. But rather than developing an abstract or global environmental philosophy to justify its existence, as was the case in other environmental movements, it remained a basically conservation oriented local group with strong local support. This is due to the fact that movement participants are mostly retired people. Though the movement became known outside its locality, but mainly within Moscow, it had no ambition to make contacts with Russian or international partners. Although individually many of the activists became interested in political careers, the movement itself remained neutral in party political terms, accommodating a wide range of political beliefs from communists to democrats.

Other Local Movements

There are a number of other local environmental movements in Moscow, apart from Bitsa, some with concrete goals and others with more general aims. Few of them are successful in their efforts.

One neighbourhood group protested against plans to build a bakery on a site formerly used for nuclear waste. They were convinced that the bread baked in the new factory would be contaminated. The fight went on from 1983 but the construction of the bakery started in 1991. The movement's only achievement was that it raised public awareness.

Another environmental group was organized against the Northern Thermo electric power station. This was part of a larger project to create an industrial zone in the northern part of Moscow with many new enterprises.

The movement organizers collected 300,000 signatures and held demonstrations (in 1989) with several thousand participants. But the construction went ahead. In fact, if anything, they felt that the protest speeded up the completion of the project. The media coverage strongly supported the building of the power station and portrayed the movement participants as "not in my backyard" protestors. They claim that they are a neighbourhood environmental group with several local concerns, like the local river, the Yauza, which was used for dumping dirty snow collected from Moscow's streets during the winter period, or trying to find solutions to the problem of garages being built all over the neighbourhood to accommodate the growing number of car owners and to combat the similarly growing number of crimes.

Many local groups have general environmental aims rather than concrete goals. One of them, which started in 1985, is located in Cosino, a new district on the outskirts of Moscow, which used to be a quiet nature reserve area just outside the capital. The reason for joining Cosino to Moscow was to find land for new high-rise blocks of flats. The local protest group of 200 people joined forces to prevent this. They also wanted to save two local lakes, which would have dried out if the construction had gone ahead, as planned. The movement is called "Ecopolis Cosino" following the teachings of a Russian ecologist, Kavtaradze, who talks about ecopolises, ecologically sound places, as opposed to megalopolises, large cities with a deteriorating environment. The leader of Ecopolis Cosino is a strong believer in Kavtaradze's teachings.

The utopian ideas of an ecopolis and the concrete goals of protecting the neighbourhood from becoming a victim of the ever growing Moscow estates are combined in the main aims of the movement. They do not organize mass demonstrations, but prefer lobbying individuals in charge of decision-making up to the level of Yeltsin and are very proud that foreign journalists or ecologists frequently visit them and report about them. They have achieved some of their aims: the plans were revised and fewer blocks were built than first planned. The idea of letting a small area of Moscow remain an "ecopolis" was accepted by some of those in charge within the authorities.

There are several local environmental movements in Moscow which are concerned with general environmental problems in the neighbourhood rather than one particular objective. The neighbourhood Grassroots Environmental Movement in the Leningradsky District of Moscow, the environmental group in the neighbourhood of Fili, and in Strogino, and the group in Lyublinski district of Moscow are examples of this. Their aim is to prevent diseases occurring as a result of contamination, to obtain proper information about the situation and its possible consequences, to persuade firms to use better filtering and purification methods and to introduce environmentally safer technologies. They seek to put pressure on the authorities to introduce and implement proper legislation. These districts are not necessarily the worst in environmental terms. As one of the activists argued:

> Movements do not necessarily appear in polluted districts. Interestingly enough, ecological movements are developed best of all in those districts where some green areas have been preserved. Movements do not always appear as a reaction to pollution, people react sometimes to relative changes. (Zaykonova, p.2)

At the demonstrations, which were frequent in the late 1980s and early 1990s, but have since become rare, they used to have hundreds or even thousands of people protesting. Today the movements can count on no more than fifty participants in any action. The core activists usually number around ten. The most interesting characteristic of these local movements is that they are mostly middle aged or older people, rather than members of the younger generation. Often it is a cross-section of the population in terms of occupation, retired people or housewives, though the leaders are mostly well educated with degrees in geology, geophysics, physics, biochemistry, for example. The majority of the leaders are women. Most of the local movements belong to the Moscow Federation and recount active support from the MEF, which they appreciate very much. Most of them have very few resources other than their own enthusiasm, and mobilize all the work-related help they can, such as photocopying, to save on expenses. As local phone calls are still free in Moscow, it is a matter of time rather than money to maintain contact via the phone, which is why telephones are "manned" by pensioners most of the time. All the local movements we contacted put special emphasis on educating Russian people, changing their priorities from the economy to more long term and global problems, such as the environment.

Often movements developed from a previous Russian custom of letter-writing to communist party committees complaining about things which concerned them in the neighbourhood. When the glasnost period allowed them to unite and organize themselves without police repression they met up on a regular basis. The political atmosphere has changed but the authorities have not. And environmental issues were politically safe to unite people of diverse political affiliations. "From the beginning our motives were political, but to attract the public and create a social base we decided to use an environmental movement," argued an activist (Shalimov, p.3.).

Thus the movements do not achieve much in concrete terms. Occasionally they might stop a previously green area from being turned into a waste dump but generally their main concern is the lack of legal protection against industrial pollution, and the lack of information about the level of pollution. Often they repeat hair-raising anecdotes about nuclear contamination which killed one of their neighbours, or sudden 25 per cent rises in child mortality rates as a result of the mercury level, as well as the alleged murders of green sympathizers who refused to sign documents allowing more industrial plants to be built in Moscow, and so on. These anecdotes are difficult to verify but have strong psychological effects among those who feel they have to act upon them. The authorities usually deny these allegations, but when they are in dispute over concrete plans, the argument often ends in disagreement over priorities. The argument often turns on whether it should be the economy and people's every day needs, and living standards which should be given priority, or more long-term, environmental aspects. Those in the movements at local or national level are very much in a minority but they certainly attract the support of those who support green priorities.

Concluding this part of the chapter, we have introduced several Russian environmental movements. As in the Hungarian case, they were examples selected as typical of the environmental movements which exist in Russia. Thus federal organizations have been described here as well as different kinds of local movement. We now turn to an evaluation of Russian environmental movements.

An Analysis of the Environmental Movements in Russia

Generally speaking the environmental movements in Russia which were born in the late 1980s went through two major periods. The late 1980s and very early 1990s witnessed a newly-found freedom which allowed people to express their discontent and demands by demonstrating. This was not unknown in other former socialist countries in the early period of their democracy either but in Russia, on the one hand, negative feelings towards "deviant behaviour" and "hooliganism" was stronger than in Eastern Europe and, on the other hand, participation in (state organized) mass demonstrations, such as 7 November celebrations, were more customary and appreciated till the end of the Soviet regime. Attending demonstrations was therefore both a well known practice for expressing collective feelings and a new and welcome experience, because this time it was not state organized. Even if individualism had grown by the late 1980s, relative to earlier periods, the idea of collectivism among Soviet people remained much more important than in Eastern European socialist societies which had a shorter Communist past and felt that it was imposed on them from outside. Russians emphasize that collectivism had been a cultural imperative historically. Hence collectivism is not regarded as a foreign concept which was imposed on them, as in Hungary and other Central-Eastern European countries. The numerous and well attended demonstrations and strikes of the late 1980s and early 1990s, frequently used by environmental movements as well, were not, however, welcomed by the newly forming Yeltsin regime and ideological propaganda turned against them. The tragic events during the two coups, resulting in unnecessary bloodshed, further strengthened the official political view that demonstrations are dangerous and irresponsible political actions and should not be used by civil initiatives. This message was well accepted by most environmental movements and in their second stage of "development" all of them started to denounce demonstrations as a method of political action and abandoned them.

The other important characteristic of Russian social, including environmental, movements is not entirely independent from the above described attachment to collectivism. Social movements in Russia, once organized at grassroots level, felt a strong need to combine forces. City and federal level umbrella organizations appeared soon after local movements were organized. This umbrella system always has a hierarchical structure: the federal, national or city level incorporates and claims to represent member organizations at local levels. In the case of environmental movements both the Socio-Ecological Union and the Moscow Ecological Federation represent this tendency. Local movements, on the other hand, confine their activities to the locality which gave birth to them and rarely gain a national reputation. This, as we demonstrated, was very different from Hungary where local movements did become nationally known. In Russia local movements feel weak and the umbrella organization provides them with the necessary backing in confrontations with the authorities. Hence federations are very much welcomed by local movements. Almost all environmental movements came into existence during the late Gorbachev period, along with all the other civil initiatives of the time. The exception was Greenpeace of Russia which was not a grassroots movement.

The participants

As in Hungary, the participants in Russian environmental movements are often people with a high level of education, mostly with degrees in natural sciences. In local movements, however, usually only core members are educated and a wide variety of "foot soldiers" participate in the movements, from ex-army officers to factory workers. People with higher education are particularly welcomed not only because of the special knowledge they can contribute to the movements' activities. The intelligentsia in Russia have always played a special role in society. Back in the pre-revolution period they were the most progressive independent thinkers in a generally very conservative environment as far as social change was concerned. More recently the lack of a western type of middle class provided them with a special role in society within which the distinction is between working class people, peasants (farm workers) and the intelligentsia, usually defined as people with a degree. Education has generally played an important role in the Soviet Union and the majority of the population benefited from this. Education was the most important channel of personal improvement and career advancement, and was a popular route to follow. Those with university degrees have a high reputation and prestige even if it does not necessarily lead to very different living standards from the rest of the society. It is not surprising therefore to see that the so-called intelligentsia is well appreciated by most Russian environmental movements.

The most important distinction between local and national movements is, however, in their age structure. Many of the participants in local movements are retired people, people on long term sick benefit and housewives. The specificity of the Russian local environmental movements therefore is that they attract an overwhelming majority of the older generation. Young people rarely participate. National movements, on the other hand, have more younger people, many of them in their thirties. Greenpeace is the exception with a majority of thirty to forty year olds among the core members. This could be connected with the fact that Greenpeace Russia was set up by Greenpeace International which itself operates with a younger generation and "imported" its idea concerning age groups into Russia. The Socio-Ecological Union, which traditionally recruited members in the natural science faculties of universities in the period when it used to be the Nature Protection Squad, also used to have many young participants. But now the Union has widened its scope it attracts a mixture of students and middle aged former students.

Environmental movements, as mentioned above, were regarded in the "transitional" period as oppositional political groups. Thus movement participants were environmentalists and oppositional political activists at the same time. Unlike in Hungary, where this phenomenon also occurred but these two groups of people, "the oppositional" and "true environmentalists", separated after a short period of time, in Russia they stayed together longer. Political parties remained fuzzier than in Hungary and it was more difficult to distinguish among them up to 1993. Elections became a regular feature of Russian political life, however, and many activists from the environmental movements became political candidates. At the 1990 elections environmental issues were a popular subject which attracted many voters and the political parties often courted environmental movements for their support. Most candidates declared their interest in environmental issues in the run up to the elections, but did not always fulfil their promises once elected. Russian participants in social movements do not distinguish as strictly between their roles as movement activists and political candidates in local or national elections as do their Hungarian counterparts. Political parties in opposition are not separated in the activists' view in the way they are in Hungary. In fact, most of the core movement members wished to be elected at some level of authority. They, and the rest of the movement participants, believed that a movement activist who becomes an elected member of the authority would give the movement power and influence. Once elected as councillors or deputats (representatives in the Parliament), however, many of them started to look at problems from a different angle to the great disappointment of their colleagues in the movements who felt let down. Activists frequently accused these former movement members of using the movement as a political springboard to gain popularity in the neighbourhood and be elected. Perhaps it is worth noting that, unlike in most other countries, elected representatives at any level are paid a salary in Russia. This led to the suspicion that being elected was also a job-seeking exercise in some cases. But not all movement participants saw their activities as a step to potential political careers. Most of them remained devoted environmentalists pursuing a matter which was not easy. When demonstrations ceased to be the most important way of protesting, their task became especially difficult and tiresome. Most movements are very poorly funded, if at all, which means that personal sacrifices were often needed to achieve their goals. Very few movements became well funded enough to pay their staff. In fact, only Greenpeace Russia and the Socio-Ecological Union were able to do so.

The structure of the movements is generally very similar to that described in Hungary. The core members can rely on a group of regular activists. The second circle was made up of people who appeared in demonstrations or later were more likely to sign petitions. The widest circle is the group of sympathizers who support the group's activity in principle. Many of the latter regularly contribute to the movements financially, if this activity is properly organized by the movement itself, as in the case of Greenpeace. Women's participation is very high among Russian environmentalists both among the core members and the regular activists.

The Leadership

The movement leaders are overwhelmingly people with higher education, as expected. As it is difficult to compare the western type of middle class with the Russian social structure, the reasoning of the western theoretical literature which argues that middle class, public sphere employees are more likely to participate in social movements, cannot be applied in those terms. But I would suggest that the so-called intelligentsia in Russia fulfils a role similar to the middle class in the west in many ways, and especially as participants in social movements, where they are in the vanguard. If we accept this argument, then it can be suggested that in the Russian circumstances the term middle class can be replaced with the equally disputable term intelligentsia, in which case we can and should argue that environmental leadership is most certainly attracted by the most advanced group of the society in intellectual terms, just as in western societies.

The leadership of environmental movements in Russia was hierarchical in its practice in every case. This is not surprising. The Russian tradition in favour of hierarchical organization is very powerful. Although there were strong signs of rejection of hierarchical structures in principle, in practice this led only to the concealing of the hierarchical structure. For example, instead of naming certain functions "leaders", they called them "coordinators". This did not solve the problem, because in practice the so-called coordinators were clearly leaders of the movement. In line with the popular democratic ideas the movements also tried to introduce "proper" representation within the umbrella organizations, which also turned out to be futile. Wherever such councils or assemblies were set up they failed to function. Member organizations did not wait for formal (ir)regular meetings to express their views. Indirect contacts with the umbrella organizations' leaders turned out to be more useful and workable than formal representation.

Bureaucracy is the other traditionally well developed form of organization in Russia. Like hierarchical structure, bureaucracy was also rejected by movement participants, but this time not only in principle. Bureaucracy was looked upon as a suffocating evil of society. As a result many environmental movements refused to introduce "bureaucracy" by not documenting their activities. This clearly hindered them in their activities, as was recognized occasionally but no solution was found.

The Movements' Goals

In terms of their goals, there were three types of environmental movements in Russia. Some came into existence because they wanted to achieve a concrete goal. These were in the minority. The movement for the protection of Bitsa forest for example was one, which had a very concrete aim in 1987, to stop the project to build a zoo. Having achieved this, they became concerned with the whole forest and remained a movement with this relatively narrow aspiration.

Secondly, a number of environmental movements, both at the national and local level in Russia, were triggered by a concrete event and later adopted general environmentalist goals. The original concrete aims included wanting to stop river diversions (SEU), stopping the construction of building projects (Ecopolis) and thermo electric power stations (environmental group North) or a bakery on a former nuclear waste site (Khoroshevsky district movement). Such movements converted themselves in their second stage to becoming general environmental movements with wide concerns.

The third category of Russian environmental movement was organized without any concrete goal or triggering event other than the general political context. These movements, which were again both local and national, had a very general environmentalist agenda from the very beginning of their existence. They were concerned with general health hazards in the neighbourhood, industrialization, growing construction works in an overcrowded city, cutting down trees, polluted rivers, lakes and air, the growth of garage building, noise from motorways, and so on. (Specific examples are: the Strogino neighbourhood movement, Fili environmental group, Leningradsky district environmental movement, Lyubinsky and Moscow Ecological Federation environmental groups and Greenpeace Russia).

Whether the environmental movement came into existence as a result of a concrete goal or was based on a general purpose concerning their environment did not make any difference to their survival, unlike in Hungary. Concrete goals did not have to be widened to "global" environmental concerns or even into general ones to secure the durability for a movement, as we can see in the example of Bitsa. It was sufficient for them to broaden their interest from the zoo to the whole forest to remain a movement, once the zoo problem itself had been solved. Those many movements which originally had a concrete aim and then became environmentalist in a broader sense did not do so with a conscious aim to survive, in contrast to Hungary. Their new interest grew as a natural process from the first stage to the second one. Groups which had general aims in mind from the start also failed to think about survival strategies and acted more instinctively. As mentioned earlier, in the case of local movements the average age is fairly high (mainly retired people), which might explain why these movement participants do not think in terms of long-term and sophisticated survival strategies. Thus the empirical evidence of the Russian social movements clearly shows that the argument of the resource mobilization approach which claims that rationality plays an important role in movement organization should be disputed.

Conflicts

Russian movements are not immune from internal conflicts. Six types of conflicts can be distinguished. Firstly, there are internal conflicts which are based on personality clashes. Strong core members become involved in conflict with each other which can lead to some members leaving the movement altogether:

> Conflicts do take place. The reason is firstly my awful character and in other cases people who want us to carry out their ideas...intellectuals think that they are the only ideologists and the rest of the people are idiots. I am an academic myself and have plenty of ideas, but they do not have to be imposed on each other. This sort of conflict makes people leave the movement. (Ecopolis, p.9.)

Secondly, conflicts are based on changing priorities. For example, the problem of garage building seems to be a source of conflict among movement members as well as between the movements and the population. Some people who previously supported environmentalist ideas changed their minds and left the movement when it turned against garage building projects. The population at large is clearly divided between car owners demanding garages and those who are against it. Those in favour of car protection turned against environmentalists with whom they might have sympathised in the past. This seems to be a unique but in a bizarre way central source of conflict among Russians, appearing in several local movements (Bitsa, North, Strogino).

The third type of conflict which divides people according to their personal status is over whether they are Moscow residents with satisfactory housing conditions or not. Although in the country and especially in the cities many people feel unsatisfied with their accommodation, some environmental movements strongly oppose any further housing construction (Ecopolis, Moscow Federation). This of course again creates divisions between those who are in need of new housing and those who do not want the city to grow any further.

Fourthly, conflicts developed with industrial enterprises. Some movements are against them generally, while others oppose specific cases. Many of the movements believe that moving industry out of major cities is the solution to Russia's environmental problems. Whether these movements think on a general urban scale or on a more concrete level they still fall, in my opinion, into the "not in my backyard" type of movement (Moscow Federation, North, Bitsa, Ecopolis, anti-bakery movement). Moving heavily polluting industry from one part of the country to another is not the solution to the problem. To be fair, however, I should add that in many cases this is only part of what the movement stands for. Nevertheless it puts a question mark over their interpretation of the frequently repeated slogan "think globally, act locally", since only the second half of it appears to be kept in mind. There were, however, quite a number of environmental movements which did not confine their thinking to a "not-in-my backyard" solution (Socio-economic Union, Greenpeace, Lyublinski, Leningrdasky, Strogino, Fili).

The fifth type of conflict is related to political affiliations. In terms of political sympathy it was always easy to distinguish between pro-communist and pro-democratic tendencies which had clearly contrasting views from the first election period under Gorbachev when political parties had not yet been formed. Consequently people could develop distinctive political priorities which could lead to disagreements and insoluble political disputes and disagreement within a movement. These two kinds of political debate, the pro-communist versus pro-democratic axis, separated people in terms of their political views but did not separate them within the environmental movements. Some of the activists were communists and were strong supporters of the Soviet regime. They had strong nostalgic feelings towards "the good old days" which had disappeared in Russia. The present political and economic disorder disturbed and disappointed them. But that did not prevent them from cooperating within the same movement with people whose views were very much on the opposite side and were supporting the so-called democratic line in political terms.

A more recent development in the Russian political scene is the fairly high proportion of followers of fascist views. This clearly appeared in the debate among Russian environmentalists as well. According to the movement activists there are strong tendencies among some environmental groups to identify "cleansing" the environment with social cleansing, openly advocating racist views. The St. Petersburg branch of environmentalists and Chelyabinsk (a regional centre in the Urals) were identified as eco-fascists, and even formed a party, called the Green Party of Russia. This was a break-away party from the League of Green Party which is an "eco-anarchist and eco-socialist" party, according to Vladimir Damie, a leading figure of the group. The League itself originates from the Green Party and was founded in May 1991. The League labels itself as "eco-anarchist and eco-socialist" because they were against the social and economic policy of the government which they argued was bureaucratic and were dissatisfied with the system of presidential democracy in Russia. The eco-fascist Green Party of Russia attracted the majority of the board of coordinator members of the left wing League of Green Party when they run into serious disputes and separated.

Political diversity, although different in Hungary, was similar in the sense that it was both a principle and a practice in the environmental movements without exception. Views of eco-fascists, however, were completely unacceptable to all the environmental movements studied. The movements were not chosen deliberately on this basis but all the movements interviewed turned out to be strongly and categorically against eco-fascists. Only one of the movements, the Socio-Ecological Union, reported that they did have contacts with those supporting the Pamiat movement, which is also a racist, right wing political organization. Apart from this the environmental movements we come across expressly closed their ranks against eco-fascism.

The final conflict which should be mentioned is related to the lack of a proper legal system in Russia. Soviet law basically identified anyone who did not abide by it as a criminal case. The concept of civil law did not develop in Russia in the Soviet period and has not been established even now. Lawyers and judges have never been trained to deal with such cases or even to look at them from a "non-criminal" point of view. Neither are they trained as environmental lawyers. The idea of a politically independent judiciary is also lacking in Russia. Courts, judges and prosecutors are strongly controlled by the state. This is fundamentally different from Hungary. Advocates of the political opportunity structure approach (Kitschelt, 1986; Kriesi, 1991) and civil society theorists (Habermas, 1992; Keane, 1988) put a special emphasis on the question of an independent judiciary, arguing that this provides a very important basis of a democratic regime.

Social movements, including environmental groups, in Russia also recognize a very strong need to turn to litigation within the framework of civil law, to be able to sue those companies which breach existing regulations, but there is no possibility of doing this at present. Within the framework of criminal law the state could do something but does not. This makes environmental movements feel very weak in disputes with enterprises which are in a strong position anyway. In addition to this even the existing law is badly defined, and is full of loopholes and ambiguities. Decrees are issued on a frequent basis, often by the president, which are ignored mostly because they contradict existing ones. The highly unsatisfactory state of legislation and lack of a proper legal framework makes environmental movements feel helpless in many ways, a problem which was often raised by them as well as by the authorities (see chapter 8).

Success

Success is not a very frequent experience among the Russian environmental movements. Among those with concrete goals only Bitsa fully achieved its objective, when it managed to have the proposed zoo project abandoned and the forest declared a national heritage park. Apart from this a few movements achieved partial successes. Greenpeace was successful in fighting against a foreign company's deforestation in the far eastern part of Russia. Activists who later became founders of the Socio-Ecological Union succeeded in their protest against the plan to divert

Siberian rivers to feed southern cotton production. All these achievements, however, happened in the late 1980s except for that of Greenpeace's in 1991. Since then most movements have experienced failure in their efforts.

Some of the victories were more likely the results of economic-led decisions rather than due to the environmentalists' protest. The cancellation of the plan to build a new power station in Moscow's northern district to provide electricity for new industrial development is an example of such a case. The protest against the power station coincided with changing plans due to the economic constraints which was probably the real reason behind the decision not to go ahead with construction.

The weakness of environmentalists when fighting against powerful economic interests, such as new enterprise construction, is well illustrated in the case of the protest against a bakery. A lengthy battle by the protesting locals, who were aware that the site had previously been used for dumping nuclear waste, ended in complete defeat. The bakery has been completed and produces bread against the strong wishes of environmental protestors even though the fact that the site was used for nuclear waste has never been disputed by anyone. Economic aspects simply became dominant over worries about environmental issues.

This situation might explain why Russian environmental movements have often developed a broad rather than a concrete approach. Raising general awareness concerning health hazards seemed to offer more chance of "success" than winning concrete disputes. Talking about the condition of the air, the rivers and the lakes earns many people's sympathy. A fight against a concrete factory turns out to be futile in too many cases to encourage followers. This is why arguments of the cognitive approach (Jamison, Eyerman and Cramer, 1991), relating to the desire of environmental groups to spread green views, are very relevant in this case. Success of this kind can be achieved by the movements in very broad terms, by raising environmental consciousness. However, the priority given to problems with the economy and declining living standards are still emphasised not only by those who are in power to make decisions on the matter but by ordinary people as well. Nevertheless there is undoubtedly an increasing recognition of problems not acknowledged earlier and the efforts of environmental movements have certainly contributed to this process.

Another important aspect of the movements' achievements, as in Hungary, is the fact that people are ready to stand up for their views in an organized way when they disagree with authorities or enterprises. Whether they end up winning the fight or not, they have accomplished a new political experience which on its own is a success.

Finally, it should be pointed out that movement activists, by participating in the group, have gained socially as well. They have become a strong circle of people with a lot in common which has brought them very close to each other in many cases. Whether they were old age pensioners or young intellectuals they have all done something which meant a lot to them and created a basis of friendship and care for each other. They were doing something morally very desirable which gave them enormous satisfaction. This was most definitely one of their successes.

The Media

The media do not play a very positive role in helping environmental movements in their efforts in Russia. During the upsurge of social movements in the glasnost period the media showed a tremendous interest in reporting about environmental movements, supporting them in their protest, and thus contributing to their growing popularity and consequently to the pressure they could exert. After glasnost the routine of media control, well exercised in Brezhnevian times, came back into the political arsenal and reports on protests became scarce. Current environmental movements in Russia find it very difficult to get reported in the press, the television or radio on national level. Local papers sometimes show more interest in local matters, but their ability to give publicity is so much less that it cannot be compared with national media. Many movements pointed out that the international media reports their activities more often than the Russian. This shows that there is enough to report about, and that it is the system of self-censorship within the media concerning protest activities which prevents the Russian press from informing the public about environmental protests.

Because movements lose out on publicity, they do not become as well known as they otherwise would. The loss of numbers of sympathizers as well as the declining success rate is closely correlated with the changing attitude of the media. In the long term even public consciousness will suffer. The fairly recent process of raising people's awareness concerning environmental issues and all the efforts to change people's attitudes and priorities so that economic decisions would not contradict environmentally sound objectives will never be achieved by the efforts of any number of environmental movements alone. Without the active participation of the media this will be a lost battle in the long run, as new social movement theorists have rightly emphasised. Unless the situation changes, Russian environmentalists have an immense task on their hands.

Conclusion

To summarize our analysis of environmental movements in Russia, it should be noted that although love of nature goes back a long way it was based on a romantic patriotic notion which has only recently changed. Today one has to recognize the enormous difficulties the movements face. Public opinion supported both environmentalism and protest activities in the late Gorbachev period when many movements were born as modern protest groups. At this time frequent and well attended demonstrations and media support ensured that Russian environmental movements had some acknowledged political influence. At this stage political opposition was combined with environmentalist goals. Later, movement tactics had to change: mass demonstrations were abandoned as irresponsible political activities. When the media started to neglect them the movements' situation further worsened. As a result not many environmental movements managed to achieve successes to boost morale and reactivate participation. The lack of funding also hinders their survival. Many movements are engaged in wider issues rather then being bogged down in concrete battles which, as earlier examples showed, were intensely difficult to win.

Russian environmental movements, unlike those in Hungary, form federations in a hierarchical system. Since these umbrella organizations give them full support, information, experience and advice this seems to be a useful tactic in difficult circumstances. These federations are made up of the best educated and experienced leaders and activists who are usually much younger than most local group participants. However, there is a high proportion of women among all movements, local or federal, and women generally play an important role in Russian environmental movements at every level including in leadership positions. Some local movements were organized originally to achieve concrete goals but many of them aim at general environmental issues, pointing out problems in broader terms. Their lack of success undoubtedly hinders environmental movements in attracting mass sympathy and impedes their survival. The lack of proper legal facilities is also a very serious obstacle. Thus Russian environmental movements put up a tremendous fight in very difficult circumstances. After the initial period when they were well accepted and appreciated the political situation has changed and turned against them.

In chapter 8 we shall look at the way local and national authorities developed in Russia and examine the relationship between environmental movements and these authorities. But first we turn to the Hungarian authorities.

Note

1. Considering that Greenpeace International is not in a desperate financial situation such an extreme method seems a little surprising. Perhaps this was only a badly marketed attempt to gain local contacts before the local branch was set up.

7

Local and National Authorities Versus Environmental Movements: the Hungarian Case

Introduction

In the previous two chapters I described and then analyzed Hungarian and Russian environmental movements and, with the exception of the media, the "units" of my investigation were the movements. However, limiting an analysis to the movements, the "challengers", as Tilly (1978) called them, without examining those whom they "challenge", that is, the national and local governments, would be one-sided and unsatisfactory. Social movements are embedded within society. Hence it is crucial to look beyond the movement in order to understand why they do what they do. Most theoretical approaches to social movements are fully aware of the importance of the social context, except for the organizational entrepreneurial group of the resource mobilization school, which treats social movements as if they existed in a social "vacuum". Like most students of social movement, I argue that they are a *social phenomenon* and have to be seen in the *social* context, not in isolation. In the case of social movements it is the local and national governments which provide an important context which should be looked at in relation to social movement activities. This will be the aim of chapters 7 and 8.

In these two chapters I will analyze Hungarian and Russian authorities. The focus of the analysis will be the development of local and national authorities in the two countries since the "democratic" regime was introduced, in order to establish their attitude towards environmental movements, which will be the discussed in the second part of these two chapters.

Thus in this chapter I will first discuss the different levels of authorities in Hungary, with special attention to those which deal with environmental problems including elected and non-elected members of both national and local authorities. Then I will analyze the environmental movements' perception of the authorities and their relationship to them. Chapter 8 will follow the same pattern of analysis in the Russian context. The following chapter (chapter 9) will undertake the comparative analysis of the Hungarian and Russian cases.

The modern state is characterized by power and its legitimate use with the help of an administrative organization through which it maintains its day-to-day existence. This administrative organization plays a very important role in the life of a society. Governments decide the political direction of the country and authorities implement it. Consequently the relationship between government, administrative bodies, and social movements is crucial. It is the government and the administrative authorities which social movements face when trying to achieve their objectives. The central administration is capable of initiating changes in the legislation, modifying regulations and enforcing them. Therefore even when the target of an environmental movement is, for example, an enterprise, it is the authority (central or local) which initiates the legislation or regulation that the enterprise has to comply with. It is the (local or national) authority which is in charge of decision-making in a dispute (the legal system is part of the "national authority" in this sense), and this is the case in most Eastern European societies no less than in the West.

Consequently, whether an environmental or any social movement achieves its goals depends a lot on the situation dictated or influenced by the national and/or local authorities' initial attitude, towards grassroots initiatives. Governments, and other central or local authorities, can be supportive politically or even financially, by providing funds for the movements, or be seriously obstructive. In these two chapters I will examine the existence or lack of existence of such roles of authorities in Hungary and Russia, the development of institutions whose role is to safeguard democracy, such as government, national and local authorities, and elected representatives and the role of non-elected officials, all of which are very different in these two post-socialist societies.

Government, National and Local Authorities

National Level

The Government in Hungary results from national elections held every four years. The first free national elections were held in March 1990 and the most recent one (at the time of writing) in May 1994. There is a tradition of coalitions being formed, much as in Germany, Austria or the Netherlands and coalitions are viewed as a more successful form of government than single-party rule. During the first national election the results did not produce an overall majority for any one party, so it was unavoidable that a coalition government would be formed. The Hungarian Democratic Forum (Magyar Demokrata Forum) became the largest party in parliament, but lacking an overall majority, governed in coalition with two other minor right-wing parties, the Smallholders and the Christian Democrats.

In 1994, on the other hand, the same electoral system resulted in a landslide victory for the Socialist Party. Consequently it did not need to join in coalition to form a government but, for political reasons, decided to share responsibility and joined with the Free Democrats (Szabad Demokraták Szövetsége, SzDSz) in forming a coalition government. The SzDSz was one of the important parties in opposition between 1990-1994 and was willing to ally with the Socialists. The government since 1994 has thus been "socio-liberal". The prime minister is Socialist and only three ministers (secretaries of state) are Free Democrats, including the Minister of Internal Affairs, which is a key post, and the Minister of Culture and Education. The President of Hungary is also a Free Democrat. The reason for that, however, is different.

The role of the Hungarian president is so different from the Russian one that it is necessary to explain it in order to understand the contrast between them. The Hungarian president, Árpád Göncz, became over the years the most (consistently) popular politician in Hungary (according to regular opinion polls conducted by the Median agency). The president is elected in an independent process outside the national and local elections. The reason Árpád Göncz became president goes back to 1990. During the first Hungarian national elections after the regime change, there were two major competing parties running neck and neck: the Hungarian Democratic Forum (MDF) and the Free Democrats. When the MDF won the national elections with a very small majority over the SzDSz, it was felt fair to allow the president to be chosen from the most important opposition party as the other most important political figure of the country, the prime minister, is always a member of the winning party. Hence in 1990 the president was a Free Democrat, Árpád Göncz. His personal popularity secured him the position and has been re-elected since. In 1998 he is still the president of Hungary. The process of presidential elections, however, is now under consideration in Hungary. President Göncz remains the most popular politician but the role of president is not sufficiently safeguarded from a democratic point of view. Currently, there is no limit to the number of terms a president can remain in office. Göncz's personality, however, does not give rise to change. One reason why president Göncz became so popular among the Hungarian electorate is that, in a way similar to that of the speaker of the House of Commons in Britain, he plays a politically impartial, arbitrator role. His role is to intervene when it is absolutely necessary and not from a party political point of view, but staying above it. Árpád Göncz has only intervened in cases like the political persecution of the presidents of the Hungarian National (public) Television and the Hungarian (public) Radio by the Hungarian Democratic Forum while in government, which both provoked a strong outcry. They were both well known, critical sociologists prior to the regime change and were accused of being "liberal-bolsheviks" by the MDF prime minister of the time, Jozsef Antall. The criticism, in fact, aimed at the two media-presidents for not being "efficient enough" in (re)introducing political censorship against the growing criticism, in the different radio and television programs, of the right wing government of 1990-1994. When the Hungarian president refused to sack them (formally it was his role, analogous to that of the monarch in Britain) President Árpád Göncz earned a reputation as an exceptionally decent politician among the Hungarian public and this popularity has only increased over time. He shaped his own job from being a formal and ceremonial figure into a reliable political arbitrator who is above partisan aspects.

The highest legal body in Hungary, as in all democracies, is Parliament, where both formally and in substance (unlike in Russia) all legislative decisions are made, based on extensive political debates. Apart from that there are Subject Parliamentary Committees, drawn from all parties of parliamentary elected representatives in Hungary (equivalent to MPs in Britain, or congressmen in the USA). These committees have an important role in discussing different subjects. From our point of view it is the Environmental Committee which is the most important, because it has the power to allocate government funds for ecological projects among contesting environmental groups.

The administrative authority at national level is the Ministry of the Environment. One of its major tasks was to draw up new legislation to protect the environment. The first comprehensive law, the Law on Environmental Protection, was introduced in Hungary in 1976, under the Communist government. It was a highly unsatisfactory law, as it was not designed to be tough or efficient enough, because the state apparatus had no interest in enforcing tough measures and fines against the overwhelmingly state-owned companies. The situation today has changed rather radically. The need for new legislation, as the very first step in the process of environmental protection was well recognised from day one of the regime change by the various environmental groups, but a new law was not introduced until 1996. When, however, the new Hungarian environmental law was finally passed it was very radical compared with the old one. Thus legally Hungary did its best to be abreast of European environmental legislation. The task of implementing and safeguarding remains to be done. The government in 1991 set up twelve Regional Environmental Units, covering the whole country, which are in charge of all environmental issues in their region. They are independent government authorities, whose tasks are to monitor the level of pollution on a continuous basis in their region, to carry out measurements in their own laboratories and investigate when the level of pollution within and outside various companies exceeds the regulated level. They are also in charge of all rivers, the soil, the air, the dismantled plants, and military barracks, (including the abandoned Soviet barracks which left a very sad environmental legacy behind them).

Local Level

Local Authorities in Hungary were also reorganised as soon as the regime changed. Even their names have been changed: from council to "self-government", which was obviously cosmetic, but important substantial changes have also been introduced in their tasks and responsibilities. Firstly, local authorities are now led by an elected mayor who is the leader of the council as a well paid employee in an executive role. Thus he/she combines two major functions: leader of the elected body and chair of the executive part of the council.

The local government consists of two major parts, the elected councillors and the non-elected officers. The councillors work in subject committees, where policies are discussed, which are then ratified by the assembly of the elected councillors of the council. The mayor is supposed to create the bridge between these two parts of the local government, which were allocated considerable powers as a result of the decentralization program introduced by the new regime in 1990.

Before 1996 local authorities, which could only issue regulations and could only act against enterprises in their own territory, had limited room for manoeuvre in the absence of a nationally binding law. Local authorities, however, allocate finances for services and it is of course the local authority which is responsible for deciding on local environmental policies from waste collection to alleviating local environmental problems, by restricting building permits for potentially environmentally dangerous constructions, road-development plans or by preventing further water pollution. Local governments have considerable power since the 1990 decentralisation. Consequently for local environmental groups close contacts with local authorities are essential.

In Hungary (unlike in Russia) council officers are obliged to implement policies suggested by the subject committee after they have been ratified by the council assembly of elected representatives. The most important point here is that in Hungary in all authorities it is the elected representatives, whether MPs or councillors, who are in charge of policy-making, and officials have to obey their decisions.

Elections of local authority representatives are also held every four years, usually six months after the national elections. They are held at the same time throughout the whole country. As mentioned before, soon after the first national elections of 1990, the majority of the population turned against the MDF-led government. This was well reflected in the voting during the local elections all over Hungary. As a result, the first local elections in December 1990 brought in a large number of Free Democrat-led local authorities, often in coalition with the FIDESZ (Young Democrats), especially in urban areas, particularly in Budapest. In smaller places, especially villages, it was the so called independents who were voted in. These were mostly people who were previously council leaders and were in fact part of the progressive wing of the socialists, but at that time it felt "wise" not to stand openly as Socialists. Four years later, however, the political atmosphere has changed fundamentally in Hungary and many of these "independents" stood as Socialists in the 1994 local elections and were re-elected. This was after the Socialists became the governing party nationally (May 1994). History, however, did not repeat itself in the sense that six months after the national elections of 1994 people were not disappointed with the fairly recently elected government to the same extent as in 1990 (when they turned against the MDF-coalition). Instead, the Socialists won the majority of seats in local authorities as well, as they had in rural areas in 1990 (as so called "independents") and also nationally in May 1994.

The situation of the Free Democrats was different. They were successful in the 1990 local elections in urban areas (especially Budapest) but lost many seats in the same places in the 1994 local elections, especially in Budapest. This reflected the local populations' disappointment with the way many local authorities had been run by the SzDSz and also the bickering within the party itself, especially at local level.

Personalities, however, play an important role in Hungarian politics, as elsewhere. Charismatic figures such as Budapest's mayor, Gabor Demszky, who is also a Free Democrat, was re-elected. This, however, is a personal victory rather than a party political one, as in the case of the president of Hungary. Demszky was often judged to be the second most popular politician according to opinion polls.

Budapest city council as a whole, however, has much reduced administrative power as a result of the local government decentralisation process all over the country. The original twenty-two (now twenty-three) local district councils within the capital were allocated considerable administrative power and responsibility in the course of the local government decentralisation in 1990. Some argued that this created certain problems Budapest-wide, but it was more successful in the more than 3,100 local authorities created in 1990 nationally (Hegedüs and Tosics, 1996).

The Authorities and the Environmental Movements

In my analysis concerning the national and local authorities the key issue is what is their relation to the environmental movements. The activity and relation of the different authorities beyond their legal obligations is in question here. During the shaping of a movement the stance of the authority helping or hindering the movement greatly influenced the outcome of the movement's development. Our interviews identified several factors highlighting the authorities' responses to the movements and their choice of strategies towards the movements. The task was to uncover whether the authorities supported citizen actions in theory and it so whether this was translated into practice, for example how they reacted when the movement entered in conflict with them and put pressure on them. Did they distinguish among the different movements or did they develop a general attitude to all of them? What strategies did they develop to deal with environmental movements in particular? Were they cooperative or repressive? Were the authorities under pressure from their own superiors when dealing with environmental movements? Did they have a conflict of interest as individuals and as members of authorities? Authorities of different levels, as described above, were approached in the search for an answer to these questions as well as the movements themselves.

National Level

Since 1990 governments in Hungary, whether a Right-wing coalition led by the MDF, or a Socialist-Liberal one, have paid too little attention to the question of environmentalism. The last time political parties showed a serious interest in the subject was during the 1990 election campaign. Since then economic changes, such as privatization, and the shrinking range of social service provision have topped the political agenda. As the general politicians' perception, rightly or wrongly, is that environmentalism is not among the top priorities of the population, parties and governments do not feel obliged to even promise radical changes in the field. When directly asked, every politician acknowledges that environmental problems exist and that something should be done about them, but this is as far as political willingness goes.

As established earlier, in charge of the environment at the highest political level is the minister of environment. There have been several Ministers of the Environment since the regime change of 1990. They were all, naturally, very concerned about the attitude towards the environment within the government. They have all shown a very positive attitude towards grassroots environmental groups for two main reasons. Firstly, the Ministry of the Environment is supposed to coordinate the activities of other ministries, including the Ministry of Industry, Agriculture and Energy which obviously have many environmentally important considerations when defining policies in their own fields and the Ministry of Health and Welfare, not to mention education. The various ministers of environment, however, are not properly consulted by any of these ministers on different issues. The only way of expressing their concerns is in cabinet meetings and even then environment ministers are often ignored, as of secondary priority. The only good news is the growth of the environmental budget in Hungary from 0.7 per cent of the GDP in 1993 to 1.1 per cent in 1997; it is budgeted to rise to 2 per cent by the year 2000. (In comparison, the highest figure within the European Union is Austria's 2 per cent. Most EU countries spend less than 2 per cent of their GDP on the environment.) (Gábor, 1997)

As a result ministers of the environment ended up feeling very much let down by their own political parties which, like most political parties, declared a "strong interest" in the environment in their 1990 election manifesto, but when it came to putting political slogans into practice, gave other issues priority. This disappointment in recognising that their scope of influence was extremely limited within their own government led environment ministers to believe that social movements could be a useful political force to enhance their own political influence within the government. They felt a huge political need to put pressure on their own government and gain more support for their much neglected subject. Their attitude towards green movements was therefore very positive. The ministers wished environmental movements were much more active, and would mobilize mass support in order to reinforce their aims. They were disappointed that the movements did not manage to organize demonstrations in front of the parliament building with tens of thousands of people demanding more attention to environmental issues. The ministers also would have liked Hungarian environmental movements to unite in umbrella organisations to strengthen their own political influence which, in turn, would have improved the ministry's position within the government. But, as we have seen in chapter 5, this did not happen. Hungarian green movements were not willing to unite in hierarchically structured organizations.

Secondly, the ministers' very positive attitude towards environmental movements was based on their political convictions. They were politicians who firmly believed in political pluralism in the sense of a multi-party system, as well as in the plurality of organisations, including the rights of collective action or grassroots movements to operate outside state/party institutions. These ministers had in fact participated in environmental movements themselves before embarking on a party-political career. They were examples of those former movement activists who, as discussed earlier, became activists in environmental groups in the movements' earlier phase and later detached themselves from the movements because, as political parties developed in Hungary, they preferred party politics to social movement participation as a form of political activity for themselves. As they had also remained faithful to their interest in environmental issues, they ended up in the highest political job relating to environmental issues. Once in post, however, the ministers recognised the serious limitations to what they could achieve. Their loyalty to environmental issues combined with their disappointment in their party's political priorities, led them into an alliance with social movements as a potentially useful political force in mobilizing public opinion.

According to our investigation, other Members of Parliament, such as back-bench politicians, who were members of the Parliamentary Environmental Committee, felt very similarly. They too often came to the conclusion that the environment was a much neglected subject compared with the economy. They complained about the lack of cooperation among the different Parliamentary Committees to improve the environment as well as the lack of available finance. These MPs also felt that the Environmental Committee was in a very weak position vis-a-vis the government, which only accepted suggestions if they did not have financial consequences:

> When the government suggests a new bill and it ignores the environmental aspects it is our job to make sure it is there and is not ignored. But we can, in fact, only "smuggle" environmental points in if they have no financial consequences. (From an interview with János Fetich, a member of the Parliamentary Environmental Committee.)

Originally the first Parliament did not even want to set up this committee as all the political parties had lost interest in green matters once in Parliament. Only strong pressure from several MPs resulted in its existence. In the committee there is a strong feeling that the reason why the outdated environmental law was not replaced earlier, was because of fears that a stronger law would slow down economic growth, which is politically undesirable for the government.

Like the ministers, the committee members also feel very positively towards the environmental movements: seeing them as potential political allies who should be cherished. They too felt that a more united environmental movement, under a strong umbrella organization, would be more effective in two ways. Firstly, it could mobilize larger numbers of people which would be more effective in putting the message across to the government and secondly, it could lobby more effectively. This would be of more use for the committee in its work. The chair of the committee (in 1995) also emphasized that the committee's work benefits from the movements' "healthier" outlook, the fact that they are outside the realm of political institutions. The movements are in touch with the grassroots feelings, he argued, which is a valuable source of information for a politician. Even their criticisms were welcomed, but it was felt that the movements did not utilize the opportunities available to them well enough, meaning that the Parliamentary Environmental Committee would have welcomed more communication and advice from the movements. Committee members emphazied that political forces which lie outside the realm of party politics are invaluable in a democracy.

Committee members were also favourable to supporting environmental movements and initiatives financially, as they are directly involved in pressing the Minister of Finances to allocate money for such purposes. Finally, committee members are involved in allocating funds to the movements themselves. This way they are actively involved in supporting green movements. As money is scarce, the competition for funds is tough, and there are inevitably losers as well. Still, a fund does exist to support many environmental movements, unlike in Russia, as will be discussed in the next chapter.

Another question to look at is whether the funds, allocated by the government via the Environmental Committee, discriminate among the environmental movements. Discrimination can be on several different bases. On political criteria we have found no evidence of any discrimination. The Environmental Committee, which consists of MPs from all parliamentary political parties, has ensured political fairness. Apart from that, interestingly, none of the committee members seemed to take a partisan view when evaluating environmental groups, environmental issues did not seem to be divided by party political convictions. Thus political bias has not led to any discrimination among Hungarian green grassroots organizations.

The allocation of funds, however, has. It clearly favoured the better known movements. Those environmental groups which were lesser known had less chance of being awarded financial support. It was as if they had to prove themselves to be strong and durable organizations, before the government would give them funds. Better established movements were, therefore, more likely to be rewarded as far as government funding was concerned. This, however, also suggests that the government does not fear well established social movements and is not afraid to further strengthen their existence. Politicians in Hungary, who are in position to help social movements, are ready to do so, either because of their political beliefs, or because it is in their own interest to utilize the movements' political support, and to lobby together on environmental questions.

Civil servants in Hungary have since 1990 been required to be politically neutral. During the socialist period civil servants were, of course, expected to be loyal to the Hungarian Socialist Workers' Party (the Communist Party of Hungary). When the new regime was established, the new slogan was to change this system and separate the civil service from the government of the day. This did not, however, become the practice immediately. The new ministers got rid of the old civil service "guard" very soon after coming into power, within months. They were, however, replaced by people whom the new ministers trusted, namely those who were politically loyal to the new government. The cleansing did not stop at the old nomenclatura, but extended to those who supported any of the opposition parties. Governments are attacked for the insistence on political loyalty which is an obvious political "routine" learnt in the socialist period. Civil servants, thus, became overwhelmingly MDF supporters between 1990 and 1994 and a new change of guard had to follow after the 1994 elections (at least at the more senior levels) now specifically aiming at neutrality. Thus, it took four years for the new regime to turn into a truly democratic political system, where the civil service's task is to support the government of the day. But at least it has been achieved in Hungary by the mid 1990s. As we will see, the same cannot be argued in the Russian case.

The civil servants' opinions described here relate to the period of 1990-1995, when our primary research was conducted. Unlike ministers, civil servants had a more complex view of environmental movements. The official instruction from the Ministry of the Environment was that civil servants have to take the movements very seriously because they are part of the environmental question and the environment is an important political issue which should no longer be ignored. In reality, however, civil servants fall into two major categories in their attitudes to the green movements. Firstly, those who could be identified with one set of opinion, did not really know how to convert theory into practice or how to maintain ongoing communication and cooperation with the environmental movements, as instructed by the minister. They could see the advantages of using the "expert advice" of an environmental movement activist instead of commissioning an independent scientist, who could advise them when working on reports or briefings for the minister. They felt that the compulsory meetings with the movement activists, set up as a result of the minister's directives turned out to be futile and ended in fiasco. The movement activists and leaders complained, but the formal meetings did not lead to meaningful cooperation. The minister also directed civil servants that financial support for environmental grassroots groups should be increased substantially as opposed to other organizations. In 1991 a ten times larger fund was allocated than in 1990. But the allocation of this fund also ran into difficulties due to a lack of coordination.

The other group of civil servants held more conservative views. They were openly hostile to the movements and viewed them as "politically dangerous forces", as a bunch of people who did not understand what law and order was. They felt that environmental movements simply provoked controversy and actually hindered the ministry in obtaining political attention, finance and winning arguments within the government. They often blamed the political leadership of the ministry for being too weak when it came to decisions, for dithering and hesitating and for listening to too many sources, such as the movements. They felt that the ministry did not even have a political concept which should or could have been implemented. Their feeling was that there was no direction within the question of environmental policy in the ministry and it was argued that no one had any clear concepts of how the environment should be improved. As this group of civil servants rejected contacts with environmental movements they could not be affected by their suggestions.

Local Level

At local level two types of organisation exist, as mentioned above, the Regional Environmental Units, which are government agencies and have only non-elected officers, and the local councils which include county, city and district levels and have both elected and non-elected members. The officers' relation to social movements will be examined separately from that of the elected members.

The officers, who work for the regional government agencies or the councils, are mostly experts in the field, with degrees in natural science. They are well informed about the regulations as well as the scientific aspects of environmental issues. The regional environmental officers emphasized in our interviews that they are completely independent politically when performing their tasks, and their main concern was "expertise-oriented". Most of them were hostile or even openly antagonistic to social movements. The basis of these feelings was mainly that they regarded the movement activists as amateurs. They felt the activists were more interested in the political side of the issues, and were basically there to question and irritate. This, it was felt by the environmental agency officers, did not lead anywhere. The civil servants in the regional offices argued that the movements' role should only be educating people and not intervention in expert (i.e. their) activities, and especially not to question or criticise them. These feelings were generally shared by officers but were much stronger in the provinces than in Budapest.

Even though officers' feelings were hostile to movements, that does not mean that officers excluded them from cooperation all together. Many felt obliged to communicate with them, and most importantly, to provide them with all the information they had concerning environmental plans or problems. This is very different from the Russian case, as we will see later. The instruction to the officers to provide information to the movements always came from elected and accountable political bodies (ministers or mayors) and was complied with by the officers. Political openness in this respect was complete between environmental movements and the apparatchiks.

The elected councillors are politicians, as are parliamentary representatives at national level. It is not surprising therefore to find a much more positive attitude on their part towards environmental movements than that of the officers. Those councillors who were members of the local authorities' environmental committees always had contacts with the environmental movements, and felt that this was beneficial for their work for several reasons. Firstly, environmental movements are in contact with the electorate at grassroots level. This provides them with useful information regarding the problems people are concerned about. Social movements became centres of information and complaints. Ordinary people often have negative experiences of the authorities' response to individual complaints. Organised movements, it was felt by the councillors, can channel public concern with greater success. This is why information or even complaints could get to the councillor via a movement faster and more effectively. Recognising this, the councillors appreciated the "channeling" role of environmental movements. Secondly, unlike officers, elected councillors have no expertise in environmental issues. As most environmental movements have a concentration of activists with a degree in natural sciences and spend a considerable time investigating cases and writing reports, this was appreciated by councillors as useful professional information. Many argued that the documents, written by movement activists, were very useful sources of knowledge regarding concrete cases. Thirdly, the movements provided political support in their own lobbying activities for environmental issues as a whole or in various concrete cases. Environmental committees at local level, in the councils, are just as weak as parliamentary environmental committees at national level. Any support they could get, any pressure demonstrated by the population, organised by movements, therefore came in handy for the councillors. And finally, elected members of councils have to stand for elections every four years. It is important for them to maintain a friendly, listening image with the population at large and with the organized part of the population especially. Since movements express their views for or against local and national politicians at election time, cooperation with them is vital if councillors want to be re-elected.

Politically speaking, environmental committees were rarely divided according to party political interest. They rather offered temporary alliance with any party bloc which was willing to help in order to achieve majority votes in the assembly. Environmental issues do not divide councillors along party lines. Rather they unite in desperation at the large degree of political neglect regarding environmental issues.

The mayors are special cases, as mentioned above. They are both elected members and executive chairs of the council. On the other hand, naturally, the mayor is responsible and accountable for many issues and not only the environment. This explains why mayors are more likely to develop an ambiguous relationship with social movements. On the one hand, the movements' critical role irritates them, but as politicians they cannot be antagonistic towards them. Mayors often end up offering support to "cooperative" movements or activists while resenting militant or very critical ones. The pressure put on the mayor by "cooperative" environmental movements often persuades them to start taking action on some environmental issue, action which at election time, can be referred back to. Becoming "cooperative" with the movements, especially when they can combine forces against the central government, is what mayors are very willing to do. Cooperation with local pressure groups often is part of their political willingness. On the other hand, mayors have more than one interest to keep in mind. They obviously have to consider the interests of local businesses as well and cooperate with them too. This could create conflicts with environmental movements. Mayors are, in the first place, politicians who are concerned to be re-elected. When it is the environmental movement which is influential in the constituency, then mayors are very willing to create a "cooperative relationship" with the movements.

Environmental Movements Versus Authorities

Before the first elections in 1990 most environmental movements were hostile towards authorities at all levels. At this time authorities were not accountable. Since then the situation has changed. There is a lot of contact and communication between environmental movements and authorities at every level. As described above, politicians are especially eager to maintain contacts with the movements. As politicians dictate policies in Hungary there is communication with non-elected apparatchiks as well, both at national and local level. This has changed the movements' attitude towards the authorities considerably. The original hostility has changed for the better in most cases. The movements often find willing authority members who are ready to cooperate with them for one reason or another. Also their interests often coincide in concrete matters. Many environmental movements even receive financial support from the authorities at different levels in the form of funding or payments for occasional reports or consultancy. Their common goals leads to a willingness to cooperate on the side of the movements as well. This cooperation can be ad hoc or continuous, depending on the situation, but has become the most characteristic attitude of movements in general. It is as if both sides had come to the conclusion, that cooperation achieves more than confrontation. This does not mean, however, that the movements are less critical in their attitude, but only that they seek and find contact with authorities, discuss matters, express views. Listening to each others' arguments often leads to compromises (but not always). The movements' underlying feeling is that it is better to achieve something than nothing. There are of course cases where compromise cannot be achieved, and the conflict cannot be resolved, but most of the time there is a willingness to communicate, to talk things over from both sides. This is a very different attitude compared with the movements' members past feelings towards authorities. And even more importantly from our point of view it is different from the Russian case.

Conclusion

To sum up the relation between the environmental movements and the local and national authorities in Hungary, firstly it is important to emphasize that the most important division is between the elected and non-elected members of the different authorities. Elected members at every level are much more favourable to the movements because they see them as potential political allies in support of environmental issues. Even those elected members of local governments who might see them as trouble-makers at first, such as mayors, are ready to compromise and try to act upon the movements' pressure because they have to think of the movements' political potential at election times.

The non-elected civil servants and officials are more anti-movement in their personal feelings and see them as political activists who antagonize and are less knowledgeable than themselves, the experts. These apparatchiks, however, have no political power and have to obey the elected politicians' decisions.

This brings me to the most important question to be investigated here. Social movements operate in a political environment in which different political actors participate. Governments, national and local authorities can develop a democratic system in which accountable, regularly elected members are in charge of decision making. There can of course be another scenario in which elections are not held regularly and elected politicians have little power and accountability is not in the centre of political thinking. Hungary used to be such a society prior to 1989. Since then, however, much effort has been made by the various political parties in government to strengthen a democratic development and arrive at the first scenario. As a result, Hungarian environmental movements operate in "civilized" circumstances in which negotiations, communication and cooperation characterize the relationship between authorities and movements. They often want to achieve the same and ally. At other times it is shrewd political thinking which brings them together. Whichever is the main motive, the outcome is acceptable. Cooperation has also helped to resolve the initial lack of trust among the movement members. This cooperative and communicative character between environmental movements and authorities in Hungary is the direct consequence of the democratic development in the political system. As will be demonstrated in the next chapter, however, the situation is very different in present day Russia.

8

Local and National Authorities Versus Environmental Movements: the Russian Case

Introduction

In this chapter I will look at the development of national and local authorities in the light of the post-Soviet elections in Russia, and investigate whether the frequent reorganization of local and national governments has led to more democracy or not. In the first part of the chapter I will look at elected and non-elected members of authorities. The second part will analyze their relationship with environmental movements. Finally, I will discuss how environmental movements perceive local and national authorities' attitudes towards them.

Central Authorities

The central levels of administration and elected bodies in Russia were reorganised after the elections of 12 December 1993. These were the first elections since the Soviet Union collapsed after the coup of August 1991 and the first in the newly established post-Soviet Russia in which competing political parties took part.

Previously there were only elections with competing views but not political parties. The first elections which allowed any competition were held in the Soviet period under Gorbachev, in March 1989, for the new Soviet Congress. These were multi-candidate elections within the one-party system which was finally abolished only a year later in March 1990. These elections for the Congress of Peoples' Deputies only resulted in 400 pro-democrat representatives out of a total of 2,250 (one of whom was Boris Yeltsin) (Sakwa, 1993). The Soviet Union still existed when the next elections occurred in March 1990. There were still no political parties but an electoral bloc, called Democratic Russia, which was established in January 1990, with branches in all the major Russian towns. Thus these elections could only offer a choice between "pro-democrat" (candidates of Democratic Russia) and "pro-Communist" (the Communist party of the Soviet Union, CPSU) candidates and played a more important role at local level which will be discussed later.

In December 1993 53 per cent of the 105 million eligible electors (Sakwa, 1994) voted not only for political parties, but for a new constitution as well, which granted a special role to the president. Sixty per cent of the electorate supported Yeltsin in his wish to become a very strong president by voting for the new constitution.

Thirteen parties contested the elections to send representatives to both houses of the National Assembly: the Federal Council (the upper house, previously called the Supreme Soviet) and the Duma (the lower house, previously called the Congress of People's Deputies). In the Federal Council every unit is represented by two members. There are eighty-nine such units, and hence 178 potential delegates. The federal Council is made up of representatives of twenty-one republics (mostly former Autonomous Soviet Socialist Republics), fifty-five regions (oblasti and kraya), two special cities (Moscow and St. Petersburg with the status of regions); one autonomous territory and ten autonomous districts. All the above mentioned units have equal rights within the upper house. In terms of their political party preference there are twelve members representing the Communist Party (led by Gennadiy Zyuganov), six members belong to Gaydar's party and 145 members do not represent any political parties.

The Duma has 450 MPs (elected representatives) half of whom are elected as constituency representatives on an individual basis and half on the basis of their party list. The threshold for any party is 5 per cent of the votes. The party political elections gave the most support to the fascist Zhirinovsky's party, the Liberal Democratic Party of Russia, (24 per cent) at this election. Second came Igor Gaydar's pro-reform party, "Choice of Russia" (15 per cent). At the time of the elections Gaydar was Yeltsin's favourite deputy prime minister, but he resigned soon after the elections and his party has practically become an opposition party. The Russian Communists received 11 per cent of the votes, the Party of "Russian Women" (leader Y. Lachova) 9 per cent, the Russian Agrarian Party 8 per cent (leader M. Lapshin), the so-called Yabloko Party, a liberal-oriented party, led by the economist Yavlinskiy 7 per cent; the Russian Unity Party (led by Sachray, another deputy prime minister) which stood for moderate reforms, gained 6 per cent and the Democratic Party, led by Travkin, 5 per cent (Duka, 1993; *Heti Világgazdaság*, 1994a).

As a result of the electoral system, however, the large majority of the Zhirinovsky party did not result in their obtaining a majority in the Duma. This is because of the separate party and individual list system and the fact that in Zhirinovsky's case it was only the party which gained such a high percentage in the elections; the individual candidates were little known and were not elected. As a result the number of MPs ending up in the Duma did not closely correspond to the election results of the political parties. Zhirinovsky's party ended up with sixty-three MPs of whom only four were elected as individuals and the rest on the party list. Gaydar's party, on the other hand, could send seventy-six MPs to the Parliament, and be the largest group there because thirty-six of them were elected as individuals. The Communists have fifty-five, the Agrarian Party forty-five MPs and the Russian Women twenty-three deputies. Almost one third (131) of the MPs do not represent any political party.

The Russian government is selected partly by the prime minister of the country and partly by the president. The three most important posts, the ministers of foreign affairs, internal affairs and defence, are selected by the president. There are, however, two other, non-elected bodies, which exercise power in the Russian administration, the president's so-called advisory body and the National Security Council, or "new Politburo", as it has been nicknamed by Russians.

The president's advisory body is the government's "rival" body, according to Russian politologists (*Heti Világgazdaság*, 1994a) who claim that this is the most influential political body in the country in major decision making. This is a non-elected body, hand-picked by the president, and is unaccountable to anyone but him. At the moment it consists of seven people, but its number is not fixed. Only one of them is a minister, Pavel Grachov, the defence minister. The the other members are unelected. They include the head of the new "KGB", two military officers, the leader of the presidential administration, and the president's personal secretary (*Heti Világgazdaság*, 1994b).

The excuse of the Chechen war created an opportunity for the president to create a second non-elected body, the so-called National Security Council, which in practice took over leadership from the government. The constitution hardly mentions such a body, and its function is not clearly defined. There are thirteen members in this council at the time of writing. It is headed by Oleg Lobov, who is a member of the president's advisory body as well. The council consists of the prime minister, the leaders of the lower and upper houses and the minister of finances. These are the so-called voting members of the council. Among the non-voting members are the minister of defence and internal affairs (Home Office), the Russian nominated governor of Chechnya, the head of intelligence and counter-espionage, the deputy PM in charge of ethnic problems, the minister of foreign affairs and the minister of civil defence (Vida, 1995).

Thus in Russia the overwhelming majority of the Federal Council and one third of the Duma consist of elected members who do not represent any political party. The electoral system is also devised not to ensure that party political representation reflects people's voting within the two houses. The president has enormous personal power, and added to that he relies on unelected bodies when forming his policies.

The Local Authorities

There have been frequent reorganisations of the local government system in the last five years. Political reforms transferring political power from the Communist Party to the state system were initiated in 1988 under Gorbachev. The local elections of March 1990 were held to fulfil this reform idea by electing new representatives to the councils but they did not introduce structural changes.

In the 1990 elections of local authorities, as in the general elections, the main political difference between the candidates was between those who supported reforms, who stood for Democratic Russia, and those who did not. But even this fuzzy distinction only became significant in major cities such as Moscow and St. Petersburg, where democrats were elected to form the majority on local councils. In most councils outside these two cities, and especially in rural areas, the old guard, local party leaders, factory and farm managers entered and won the elections, that is people who were previously closely associated with or part of the nomenklatura. In Russia there was a tradition of nomination of candidates by labour collectives, rather than residential meetings. A further obstacle to nomination via residential meetings which discriminated against candidates was that they had to produce a quorum (150 in Moscow) while no lower size limit was stipulated for labour collectives. The same law obviously disadvantaged the "neformaly", the embryonic movements and future political parties, which had not been formally registered and did not therefore count as public organisations (Boyce, 1993).

The main reason for reformers not performing well in the local elections, it was argued, was not because outside big cities people had a different opinion about the old nomenklatura, but because the democrats were largely inexperienced in organising effective election campaigns, while the old guard were more experienced organisers and successfully mobilized techniques for surviving electoral challenges (Hanson, 1993). In Moscow, however, which was the vanguard of perestroika, and where the CPSU's campaign was disorganised and half-hearted, a democratic victory was achieved and 60 per cent of the seats were won in the city soviet in 1990. Apart from that there were thirty district councils in Moscow at the time.

Structural changes came a year later, in Autumn 1991, after Popov, Moscow's mayor at the time, was given the right by Yeltsin (then the president of the Russian Federation) to determine the city's administrative structure. As a result ten large prefectures and 120 municipal districts were established within Moscow. This structure, however, was again short-lived, since it was abolished by a presidential decree in October 1993. The city council was also renamed Duma, bringing back pre-1917 names both in local and national government after the December 1993 elections.

The Relation Between the Legislature and the Unelected Executive Bodies

Gorbachev started a process of guaranteeing independence to the elected bodies of authorities to ensure that elected bodies had the power to formulate policy against the apparatchiks. This has since failed because Yeltsin does not believe in democratization as the solution of Russia"s political problems. He believes in a strong leadership and a centralized political system, hence the extensive restructuring at both national and local levels.

Another problem is that apparatchiks at every level remained more or less the same people. Ministries as well as local authorities are filled with the old guard, who firstly did not wish to give up the power they had gained and because accustomed to during the Soviet period and secondly, did not change their old routines of adopting a surly manner in treating clients. Apparatchiks were and remained badly informed even in their own fields and maintained a system which was disorganised and inefficient. They did not develop cooperation between different departments and different hierarchical levels. They continued to be part of an overstaffed and ineffective bureaucracy failing to cope with the ever increasing flow of decrees and were not backed up with adequate infrastructure such as computerization. The idea of not replacing "experienced" apparatchiks who supposedly alone possessed the "necessary professional qualifications' in fact led to the maintenance of the old system. Executives held power, government departments and central ministries as well as local authority officials dictated and kept control while elected bodies remained impotent and powerless.

The situation was not helped by the fact that many newly elected representatives were inexperienced in legislative procedures, spent too much time discussing unimportant matters, and were not clear how to discuss issues efficiently and reach decisions. The frequent reorganisation of both national and local levels of authorities in both the elected and executive part did not contribute to a successful development of a working system either. The strong hierarchy did not disappear and the lack of clarity about who was entitled to do what, both within a given level of state administration and between different levels, remained the same as before. Both at national and local level the leaders, such as the mayor at local authority level and the president at national level, were granted concentrated power, which did not help elected assemblies to develop authority and independence. Leaders either kept changing the law by the power granted them or simply ignored the law (Boyce, 1993). (The problems of the environmental law and the lack of clear regulations and implementation will be discussed later in this chapter.)

Thus "democratic" Russia remained a society where administrative power was concentrated, elected bodies were powerless, the system of hierarchy was maintained, bureaucracy was enormous and inefficient, the legal system did not serve the citizen and corruption flourished. The effect of this in the field of environmental issues is the subject of the rest of the chapter.

The Authorities and the Environmental Movements

Our aim in looking at the national and local authorities is, as in the Hungarian case, to investigate the relationship between them and the environmental movements in Russia. Based on the evidence of our interviews (methodological details are discussed in the appendix), we will focus on the questions of whether elected members of the local and national government and the apparatchiks supported the idea of citizen action or resented it both in theory and in their practice, and whether authorities developed an autonomous policy towards environmental movements and what it was like. We also studied whether the authority had any other means of knowing public opinion, or any relationship with the public at large. We investigated how they reacted to pressure from environmental movements, whether it affected them, or influenced their policy-making. It was important to find out what strategies they developed towards the movements and how these worked. It will be shown whether the personal opinion and attitudes of both employed and elected members of authorities differed from those of their own authority or whether they fully identified with the authority's attitude towards environmental movements. Finally we will discuss the relationship between environmental movements and authorities.

The National Level

Political interest in environmental issues in Russia is extremely weak. At the period when environmental movements were strong there was some increase in political interest but the growing problems in the economy and decreasing living standards swiftly pushed it aside. Formal bodies in charge of ecology have been set up at all levels, but these turned out to be more formal than substantial steps. The Russian Parliament has formed a Committee for Ecology in which MPs were supposed to deal with environmental problems and formulate new national laws and regulations.

According to a leading member of the ministry, however, members of the committee have been chosen "according to the "residual" principle (meaning MPs with the lowest prestige, not good enough for "more important" committees). "Many of the MPs in the committee did not even know the meaning of the word ecology. People, who were supposed to draft environmental legislation, were illiterate in terms of ecology." (Interview with Grakovich, p.3)

Political ignorance is coupled with administrative anarchy. The Ministry of Ecology of Russia is supposed to be in charge of the eight-nine Regional Divisions covering the country, but whilst originally the Regional Divisions were part of the ministry's responsibility, they have now become independent organisations and only have to cooperate with the ministry if they wish to do so. They are, of course, non-elected authorities. The 1,500 employees of the ministry are therefore no longer responsible for the regional level. The Ministry of Ecology of Russia was also supposed to coordinate the work of several related parliamentary committees, such as the committees for forests, hydrometeorology and geology, but these committees started to work autonomously and finally set up their own ministries. In terms of general policy, which could have guided the ministry employees in their work with the remaining committees and other related organs, no-one is clear what to do and how to do it, given the absence of any kind of ecological conception at government level which is necessary to consolidate their work. This deepens the problems of the ministry, caused by its very limited financial resources, most of which are spent on staff.

The minister of environment has a similar position within the government to that of his Hungarian counterpart: the department is in a weak political position. But this parallel situation did not lead the minister to react similarly to his colleagues in Hungary. The Russian minister does not see the environmental movements as potential political allies who can channel peoples' wish to put environmental issues higher in the political agenda of the government, and who could support him in his lobbying and make the issue more prominent politically. Instead he looks at environmental movements as a threat to him and the ministry. Environmental activists annoy him, because he sees them as people who demand information, want to know the law and call on him to implement the law which is not something the environmental minister of Russia wishes to endorse. He does not recognise any advantages in the existence of environmental movements but sees them as wholly negative and is irritated by them. Thus the idea of any cooperation is not on his agenda, but calming them is. The ministry has set up a department which was given the task not of cooperating with the movements but to act as a buffer. It is, however, a pure formality since the head of department, whose job is to provide information to the movements, commercial organisations and the public at large, admits he cannot fulfil this role:

> Theoretically speaking anyone can come to me and ask for information on this or that legislation but I can't give them such information because I am poorly informed myself. I am the head of a department, responsible for information, and I can't provide the population with information, because I don't have it myself. (Popov, p.5.)

Civil servants in the Soviet Union were obliged to be loyal to the regime. Most apparatchiks were party members and all of them had to be faithful to the party political line. The majority of existing civil servants are the same people as before since apparatchiks have not been replaced. Their political views may have altered, even if their routines have not, and they could be strongly committed to any kind of political interest. Whether they are politically neutral or strongly support any of the political parties today is not discussed. There are no attempts to create a politically neutral civil service and the problem is simply left unsolved.

The majority of civil servants are not very sympathetic towards environmental movements. Some are straightforwardly against them. This is the case even among those who themselves were once members of environmental movements. Civil servants perceive environmental movements either as groups with slender interests or as groups pursuing wild philosophical views. The groups aiming at solving concrete local problems are often looked upon as environmental movements with a narrow, selfish interest with a "speculative character", only aiming at securing the payment of compensation to activists or local residents.

"All political actions of the Greens are but "democratic schizophrenia". If we look at them carefully, we will find that the majority of them are not normal people," argued a senior civil servant, an advisor of the minister on radiation (Kuranov, p.4).

Civil servants' attitudes towards environmental movement activists are very disrespectful. The movements' leaders are accused either of only aiming at political careers, to gain personal privileges, or of wanting to receive bribes. Once elected, it was alleged, they forget about their previous environmental demands. Other movement leaders were perceived as people who only want to achieve small individual advantages, like a foreign trip. The relationship between authorities and movement leaders and their "treatment" is well illustrated in the following quotation. When asked, what strategies of interaction with environmental movements the interviewee, a senior civil servant, adopts, the response was:

> Demagogy. I have to resort to demagogic methods. When the Greens raise

environmental problems they don't aim at constructive cooperation and mutual understanding, in my opinion. Their approach is simple: this is a sore—cut it out. With such a negative approach only demagogy is a right strategy. For example, the Greens in Tomsk were protesting against a contract with the French for enrichment of uranium. How can I have a serious dialogue with people who make a fuss about a small problem but disregard a much bigger one? [the danger of uranium enrichment over gaining hard currency] I may try to convince them like children that they are wrong but I won't take them seriously. Any explanations are useless with them. The cheapest method to resolve a conflict is to deprive a movement of its leaders, for example by bribing them or offering them a trip abroad. It works perfectly. Particularly because this is what many of them want to achieve anyway. (Strigulian, p.10-11.)

Environmental issues are perceived by civil servants either as something which most people do not worry about too much or as something which they gave a low priority to. Some civil servants felt that they were not supposed to have any knowledge of public opinion and should not have any relations with the public. The public is often looked upon by them as an "underdeveloped crowd" and even the more tolerant civil servants mainly saw their role as to provide them with more information rather than listening to the public's wishes.

The public's main source of information about environmental problems is the media. But as we showed earlier the media devotes very little attention to environmental questions and they are censored. The reports are often distorted and portray one or other side's point of view depending on the journalists' own perception or interest. The media also deals only with "newsworthy" items. Using it as the main source of information regarding people's views in a central administrative organization is hardly a satisfactory solution. Some civil servants do recognize this, while others refuse to admit it.

Not all civil servants, however, are completely negative regarding environmental movements. Some see them as useful ways to educate people to improve their environmental consciousness. The movements are also useful critics, some argued, "like a wolf is necessary for a herd of sheep". But even the most positive supporters of movements among civil servants emphasized that when the greens were strong and popular, and frequently attracted masses of supporters, they only saw them as "a destructive force". And even supportive apparatchiks looked upon activists as "mentally unbalanced" people who only put pressure on the ministry to impede their work. But the ministry is "equipped" to deal with such impertinence and destruction, they said. It uses the "well-tried Russian method" to deal with complaining clients: "the traditional method: to send them around the corridors, which is enough to make them feel confused and irritated" (Grakovich, p.11).

The reason for trying to get rid of environmental movements, or at least completely ignore them rather than cooperate with them, is based on the underlying attitude of many civil servants which is well summarised by one of them: "Co-operation with them [the environmental movements] can be only compared with a dog which tries to bite your leg." (Klimov, p.18) Organizing any dialogue between the environmentalists and civil servants was considered completely futile because, as the civil servants put it, the activists' and civil servants' "levels" (sic) and views are completely incompatible. Many civil servants do not see the role of activists outside the governmental system. A group of experts is more than enough, they argued, even though the adequacy of their information regarding environmental problems is admittedly highly questionable. This aspect, however, is not brought into the equation when thinking of any roles environmental movements could fulfil. Some even honestly admitted that, although they did not like the regime before the political changes, they preferred the times when the CPSU and the KGB had a strong grip on the population and prevented social movements from existing. This worked with 97 per cent of the population, said one, and the remaining 3 per cent was successfully dealt with by repressive methods. Those were the correct ways of relating to civil society, it was strongly argued by civil servants in "democratic" Russia.

Regional Level

The Regional Committees for the Use and Protection of Natural Resources were set up recently. They are supposed to take over the functions previously held by the All-Union Nature Protection Society, and some of the functions of the State Hydro-Meteorological Committee and the Ministry of Water Supply. Officially they belong to the Ministry of Ecology of Russia as well as the regional local authority, which makes their position very difficult as the two bodies fight over them. The idea is that they will be independent of the local council and fully belong to the ministry. (They are already financed by the ministry.) The Regional Committees on the Use and Protection of Natural Resources are supposed to monitor the use of newly privatized land and protect it by placing restrictions on the new owners. This power derives from the government. They are also supposed to give "expert" opinion on issues which the existing laws and regulations do not provide for and keep watch on air pollution, surface water, waste and dumps. Specially protected environmental zones and nature reserves are also their responsibility. They have branches in every district. It is not clear for these regional committees, however, whether their role is to be an inspectorate or an agency of the (local or central) administration. The local (regional) administration sometimes sets up a rival internal department for the protection of the environment, which only increases bureaucracy and hinders cooperation, since they end up not knowing who is in charge of what. The administrative chaos is well illustrated by one employee's testament:

> Ideally the new department [of the regional authority] should be responsible for the use of natural resources, while our [the regional committee's] task would be to keep their activity within environmentally acceptable limits. But this does not happen. It seems that our status is undermined because the Ministry's status is not properly defined. The same is true for other committees, like the Sanitation and Hygiene Committee in the president's administration. As long as it had this status, it was working very efficiently, for example it could close down any industrial enterprise which violated sanitation norms. But after it had been made a part of the Health Care Ministry, it lost all its power (Kreidlin, p.6).

The Regional Environmental Committees are also very badly equipped and lack adequate equipment. "All departments have only radioactivity metres and little else" (Sokolov, p.14). They are also understaffed. Furthermore there are serious doubts about the expertise of the staff: more than half of them are former party functionaries. Whole departments were transferred to the committees when local party administrations started to be dissolved. These people are not experts in any field, except for Communist Party administration. Another group among the employees in the regional committees are not former Communist Party functionaries, but are also poorly equipped for the job. These "experts" have only secondary school education. And this in a country where there is no shortage of highly educated scientists. These committees therefore look more like places which were invented to create jobs for former party functionaries and other "reliable former comrades" rather than regional bodies in charge of environmental problems.

But staffing is not the only problem these committees face. The laws and regulations which they are supposed to follow and implement cover only 5 per cent of the problems. A major environmental law, the so-called Russian Federal Environmental Protection Law, which was issued in 1992 and was supposed to be interpreted for implementation by the Ministry of Ecology, has not been completed and has not been used in practice ever since. Old laws are frequently replaced by new ones, but not even the older ones were implemented. The newly issued ones also often contradict existing ones. This leads to a very "open" interpretation by any body concerned. For example, the law concerning fines for environmentally damaging enterprises or individuals does not work because it is too complicated and there are disputes over the meaning of passages, we were told. Or, in other cases, the court needs calculations for the size of damages, but no-one knows how to do this, so the courts do not accept the committee's application for legal action. In fact, even the prosecution's level of competence is questionable, because as one civil servant put it: "The prosecution lawyers' lack of competence can be illustrated by the fact that they phone *us* asking what laws apply to a particular problem!" (Kriedlin, p.3)

Local authorities also use another method: they "misinterpret" the law if that is in their interest, contradicting the committee's interpretation. The lack of clear guidelines from the central level allows this.

The civil servants working for the committee, however, did not develop the same antagonistic stance towards environmental movements as their colleagues at national level. They do not look at them with the negative attitude which characterizes civil servants working for the central administration. They are also aware that environmental movements were generally much more active and powerful in the past and that their influence has since weakened. The regional committees are recently established organizations, as mentioned above, and they are also known for not being very effective. This is why they have not often been approached by either national or local movement activists. Local movements do not turn to them for two reasons: they are neither local representatives nor local officials. The national movements also by-pass them and turn to the central level of administration because they are a regional body with no clear responsibilities. Since environmental movements do not challenge them, the civil servants of the regional committees do not look at the activists like confronted apparatchiks, as do those in the ministries. Committee employees also are of the opinion that environmental movements are not very effective and therefore do not see them as irritating "enemies". Rather they perceive them as independent groups which should highlight concrete environmental problems. Thus to some extent as potential "allies". When ecological movements contact them and provide them with information about various potential or existing environmental damages in the region this is accepted as valuable information, regional committee officials argued (Kriedlin, p.17). In fact, they would be grateful if the movements were monitoring a wider range of issues, such as the piling up of rubbish or dangerous waste lands, which are often not dealt adequately at present. This would encourage the committee itself to pay attention to these issues. Some of the committee officials not only are not antagonistic towards the movements but actively cooperate with them. For example, twenty to thirty students, activists of the nature protection brigade of the Moscow State University, regularly help the Moscow Regional Committee in their work on improving the Moscow region's environment. One of the former students, who had been a member of the brigade since 1985, in fact became an employee of the committee in 1991 and mobilized his "comrades" to help on a voluntary basis which was well appreciated by the committee's other employees, including senior officials.

The committee also draws information regarding the state of the environment in their region from contacts with the public. This happens mainly through complaints arriving directly from members of the public. When either an individual or a groups of people sends a letter of complaint to the ministry of the environment it passes the letter down to the regional committee in order to investigate the case, initiate action and reply. Instead, however, most of the complaints coming from the public are treated with a large degree of scepticism by the committee officials and looked upon as "overdramatized" demands. Thus, though not openly antagonistic to ecological groups, who are often behind these complaints, in fact, very little is done to carry out serious investigations as a response to them, and even less towards solving the problems highlighted by environmental groups or environmentally sensitive individuals. The letters are replied to, but only as a formality, and this, in fact, seems to be the main occupation of the staff within the committee. On occasions even the committees own employees have been ignored and their concerns are mostly disregarded. As one of them put it:

> During my working day I have to sit in my office, because the boss does not like us to go out even to the nature reserve, which is part of our responsibility. So, I have to visit it on weekends. Then I have to write a report which no-one seems to need here. It happens that I send a letter of complaint myself to the ministry and it is sent back to the committee. (Boykov, p.8)

Overall civil servants at the regional level are somewhat more positive towards environmental movements than civil servants at the central level but that does not mean that they are fully supportive. Ironically, even those who themselves participate in some environmental movements or who accept the support of voluntary helpers, are only ready to concede to those few green groups they are in direct contact with and reject the rest of them. Thus they strongly criticize environmental movements in general for various and often contradictory mistakes, such as being too global or too local, too demanding or not looking for the right things in the eyes of the official, or as people who are amateurs who should be ignored. But these civil servants at least support people with ecological thinking and are not against the *idea* of independent environmental movements on the whole. They are ready to co-exist with some of the green movements and support the "ideal" movement which they wished existed in the shape and form they envisage it. They are thus, in this sense, a lot more positive than civil servants at the central level in their dealings with environmental movements.

Local Authorities

Local authorities have elected and non-elected members but a peculiarity of the Russian situation is that local councillors, like MPs, are actually paid a full salary for their services, and in practice being an elected local representative is a full time job, with offices in the council building itself. This means that both officials and councillors are in practice employees of the local government. Nevertheless they have very different positions and outlooks within the administrative system and we must look at their relation to environmental movements separately.

The Non-Elected Members of the Local Authorities

The non-elected members of local Dumas have a very similar stance to civil servants of the regional committees concerning environmental movements. They too have surprisingly little direct contact with local environmental movements, and see them as harmless forces, as long as they are ready to compromise and eliminate extremist ideas and individuals among them. Environmental movements are perceived as groups which accommodate too many people who joined the movements because they are "odd", "antisocial" and have "extremist views". The activists, Duma officials argue, have a very one-sided approach when they see environmentalism as the most important problem. These apparatchiks believe that it is the economy which does and should have priority when it comes to allocating resources or resolving a concrete conflict. Environmentalists therefore are often perceived as people who represent very narrow interests, opposing everything the rest of the society stands for. Using a concrete example, as argued in one of the interviews, a power station should be built without further debate, if the economy needs it, and no environmental consideration should be allowed to alter or even challenge this. The solution of problems is often looked at from a financial point of view, that is people who complain should be paid compensation and/or polluting firms should be fined, and these should satisfy everyone.

Local authorities, as executive bodies, are not very concerned with environmental issues. They allocate either very few, or no resources at all to solve environmental problems and do not even have officials dealing with these issues. There was not a single official, for example, in the whole of Moscow city council, whose job was to deal with environmental problems. Consequently, there is no-one to investigate the problems signalled in the many letters individuals or groups write to the Duma concerning environmental complaints. Thus ecologists' demands are simply ignored. And this is the best they can do, officials argued, because social movements are a kind of "pest" anyway. If not ignored then they would be fought on a political basis and ridiculed. The fact that the authorities do not use dirty campaigns against the movements is considered by them to be a generous gesture, a sign of huge political tolerance. This is because "luckily" the authorities are pre-occupied with other problems. Among them is the task of maintaining political and administrative power vis-a-vis the elected members of the council. "This is a large, well-organized, highly intellectual organization which would like to preserve its power ... it is a natural inclination of people who were previously in power to keep their power," commented an apparatchik from Moscow city council (Makagonov, p.5).

And this power can and is turned into marketable commodities, such as property and privatization vouchers. The previous elite is working very hard to maintain its position using new and old methods. Environmentalism is way down the line among the worries they are concerned with.

The Elected Members of the Local Dumas

The elected members of the council are in a very difficult position. Since 1990 many councillors have been elected as new members without much experience of local government. The "old" councillors, however, also faced difficulties due to the changing division of labour between the elected and the executive body. Theoretically the councillors' work is (a) to monitor whether the law is observed, (b) prepare regulations if needed and (c) to approve the budget and monitor expenditure. The idea that councillors are supposed to make policies and decisions is, however, not clear for Russian councillors. They do initiate policy-related ideas occasionally but more often choose pet projects rather than systematically looking at different issues. But even on those few occasions when they draw up drafts, discuss them with the relevant committees, and pass them to the executive body, they are completely ignored most of the time.

Thus even concerning fundamental issues, such as budgeting, the entire elected body is completely ignored. Councils keep on operating without approved budgets, via personal negotiations between the mayor and the presidential office. This way no-one has to, even formally, approve or know about the budget allocation outside the mayor's office within the council. One year Moscow city's yearly budget, for example, had not even been discussed by September. This, according to councillors, became a practice to avoid the uncomfortable discussions which happened over the budget allocation in 1991 when cuts were introduced from the already very low allocation of 1 per cent for environmental projects to 0.7 per cent. For the following year, Moscow city did not wish to allocate any money at all. By not making the budget public, they avoided "unnecessary" debates with the councillors. Thus the elected body not only does not have the right to be in charge of the budget but is not even informed about any decision concerning major issues. Clearly the Mayor and his department keep control of all decision-making, if necessary unlawfully, and the elected members of the local authorities have only a mere formal role. They are reduced to carrying out less than their legal obligations. This means that in practice the full and part-time councillors' main work is often to reply to the public's complaints. The council is bombarded by letters from the public—from individuals as well as groups. Some are kept within the executive part of the council and replied to from there, but the majority of these letters are passed on to the councillors. They can look at the problem before replying if they wish to do so, but are not obliged to do so. Their main task is to try to reply to as many letters as possible.

Party politics also play very little role in the elected section of the council. Councillors might belong to one party or another but many of them stand for their own individual views and are elected on that basis. Party politics do change rapidly, as parties have been forming over the years, but at local level they seem to have even less importance than at national level. At the 1990 local elections, for example, even if political parties were in an embryonic shape, it was clear that the majority of the newly-elected city councillors in Moscow were anti hard line, pro-democracy members. This, however, was completely ignored when it came to appointing a chair of the elected body, Mr Gonchar, a former district party committee secretary. He was a useful choice for the Mayor, however, as Mr Gonchar agreed with the Mayor's policies and contributed to the maintenance of the old status quo. The executive body's position in the power struggle is carefully strengthened at every point. The argument is that "old" experience cannot be beaten by new ideologies.

Elected councillors do support environmental movements as long as they are moderate and do not voice too radical demands. Many of them either stood with environmental issues at the centre of their political manifestos or at least supported the subject very strongly. Even the Mayor of Moscow at the time, Popov, promised special attention to environmental issues when he was a candidate. Many of these people, from the mayor to the councillors, however, found "more important" problems to deal with once elected.

Not being negative about organised civil actions is, however, one thing, but supporting them in concrete cases is another. Some councillors even in the ecological committee, often environmental movement activists themselves, might take a conservative view, having looked at concrete complaints regarding green aspects. For example, when a neighbourhood complained against the heavy pollution caused by a tyre factory a councillor commented:

> After I studied the documents I realised, that this factory did not do any harm to the environment and did not pollute the atmosphere. I recommended that the application was turned down. This was not based on my opinion but on my professional knowledge. (Vorobiev, p.10)

And when he was confronted with the fact that thick soot and dust pollute people's houses in the neighbouring blocks of flats and that heavy industry should not be located in the middle of highly populated housing estates, the so-called green councillor, regarding himself as a "highly respected professional", a well equipped scientist to pass judgement on various matters, argued this way:

> Advanced technology always causes the pollution of the air. I can't tell exactly what problems arise, but I know that this is a natural consequence of any growing economy. One can observe this happening in the world of crystals, my academic subject. Growing crystals begin to "pollute" the surrounding environment thereby slowing down their growth because they consume the components essential for it." (Vorobiev, p.11)

The philosophy, which is very strong among most Russians, and shared even by many "environmentalist" councillors, is clear: they are convinced that the economic needs of Russia require the old-fashioned type of industrial growth, which in turn requires sacrifices. People and the state of the environment, including the air we breath in, is just one of those things which inevitably fall victim to this process. The philosophy echoes ideas repeated when the primary aim was building Communism. Human and natural sacrifices are still viewed as an inherent part of development for many people brought up on Soviet philosophy and this view persists even among environmentalist councillors as well.

There were quite a number of environmental groups in the early 1990s. In Moscow alone there were around 160 groups known by the city council in September 1992. Some were more active in creating relationships with councillors than others. There is always a number of councillors ready to listen to those movements which approach them and offer some help. They would first of all make documents available for the movements, as long as these are available to the councillors themselves, which often constitutes a problem. Secondly, these sympathetic councillors would look at their complaints and advise them about the circumstances. Some have supported a few movements with funds, even though this has been rare and the funds were not very large either. This is partly due to the general situation regarding environmental expenditure within the council.

Some councillors complained about the lack of any strong organized national or city level organizations and even offered to organize some themselves, which seems like a confusion of roles. The chair of the Environmental Committee of Moscow City Council, for example, expressed a strong wish to organize another Moscow Federation of Environmental Movements, because he was not satisfied with the strength of the existing one. He also organised a body of environmentalist councillors. This strong belief that small groups have to be organized into national or federal forces was thus not only present among the movement activists, but was very characteristic of those who strongly support environmental movements among the councillors. The lack of political strength of environmental movements was blamed on organizational matters by these Moscow city councillors who themselves were very weak politically within the city government. Their way of thinking was similar to that of Hungarian politicians who sought the backing of mass demonstrations organized by environmental movements in order to strengthen their own political power within the government.

Elected members of the council also saw the advantage of environmental movements in providing information about the state of the environment within the community. Their source of information was not confined to monitoring the press, as was that of the civil servants. Many of the councillors paid careful attention to opinion polls and felt very concerned that the media did not supply enough coverage to enable people to be properly informed and form an opinion. However, councillors themselves found it difficult to do much about this. The press, they also felt, was censored: many subjects were not discussed at all, and others were not adequately presented to alert ordinary people. The chair of the Environmental Committee himself was banned from broadcasting because of his known political views.

Councillors believe that environmental movements in Russia should consider themselves lucky not to be harassed by the apparatchiks of the authority. This is only because environmental movements are perceived as powerless organizations not worth worrying about. At the time when the movements were viewed as a stronger political force, attracting considerable public support, the authorities were not in the same position as they are now. In the late 1980s and very early 1990s Russian authorities reacted to pressure from organized grassroots action. Since then the situation has changed. The apparatchiks have regained political power, which they were accustomed to in Soviet times, and do not seem to fear social movements any more. Social movements for their part have also calmed down; they are not capable of exerting strong pressure on the authorities. Interestingly, however, many councillors argued that increased political pressure from the grassroots could even be dangerous for them, as it could bring about an unfriendly reaction by the officials of the authorities.

Environmental Movements Versus Authorities

Environmental movements are fully aware of the lack of support they receive from the authorities. They feel that their opinions and protests are ignored. Some support from councillors is experienced by the few activists who contacted them. Most, however, felt that the authority was antagonistic whenever it could be. For example, many movements tried to legalize their position by registering at the local council. In most cases this procedure became a very unpleasant experience. Instead of providing help and advice on how to draw up the statute of the movement, which was a condition of registration, council officers delayed looking at it and after a while would find some problem with it, which, the movement activists felt, was a mere excuse to avoid registering them. This often led to a lengthy process before they could finally register, and in some cases they were delayed indefinitely.

When it came to obtaining information on concrete issues and discussing them with the relevant authority, lack of support was again the most common experience. Information was not made available to movement activists, relevant data were withheld from them, and discussions seemed to be completely futile. No signs of cooperation were offered by the authorities. In the case of concrete struggles most of the time the authority ignored the environmentalists' arguments and took the side of the other partner in the debate, for example the enterprise. In the case of more general demands the movement was simply ignored altogether. Those few councillors who remained faithful to them and actively supported (moderate) environmentalists were hardly visible to the activists, as they were outnumbered by the many hostile and antagonistic members of authorities.

Conclusion

Central and local authorities in Russia show strong signs of developing in the direction of re-centralization and the restoration of non-democratic forces. This is reflected in both the frequent reorganization of the authorities as well as informally, by not letting elected bodies carry out their duties. The events of the two coups in the early 1990s, and later the war against Chechnya, provided ample and well exploited excuses for President Yeltsin to set up bodies which contradict all the efforts of the late Gorbachev period to change the authoritarian system into a more democratic one. The tentative alterations have since been overruled by many new ones since Yeltsin has become president, including at the level of local authorities.

Political parties did develop in Russia over the years, but elected representatives often do not stand for parties when they contest elections. The overwhelming majority of the upper house is made up of so-called "independents", as is one third of the lower house. At the local level political parties play even less of a part when electing councillors. As a result of the electoral system, however, political party representation would not result in corresponding representation in the elected bodies either at local or national level anyway, for the reasons given earlier.

In this situation the non-elected part of the authorities, both at national and local level, plays a disproportionately powerful role. They have successfully maintained their power to form policies and allocate budgets, and to issue decrees and regulations at all levels. Elected representatives play a mainly secondary role and feel unarmed in this struggle for power. The lack of a systematic legal system also contributes to the difficulties.

Non-elected authority members are mostly not very supportive towards environmental movements, but even some elected representatives are antagonistic towards them. The minister in charge of the environmental ministry is just as hostile towards them as the top civil servants who do not see much function for independent, non-governmental organizations and do not wish to listen to them or cooperate with them. Civil servants at the regional level have a more mixed view of environmental movements. They can see their use as unpaid volunteers, as long as the activists restrict themselves to fulfilling certain useful tasks for the authority. Radical views are strongly opposed by civil servants at all levels and many movements' complaints are viewed as a nuisance.

Local authorities have two parallel bodies in full-time jobs: elected councillors and apparatchiks. Of the two, the executives possess real decision-making power while councillors can only advise and are mostly ignored. The council apparatchiks do not have a positive attitude towards environmental movements, and generally ignore environmental problems altogether. This is demonstrated by the fact that at the city level no money was allocated to environmental issues by 1992 and that no department was in charge of environmental problems within the executive part of the local Duma. The elected body of Moscow city council, for example, has an environmental committee, but it is very weak and its activity is largely reduced to replying to letters of complaints from the public. It cannot fulfil its role in influencing financial decisions or policy making or in monitoring whether the law is observed by the council and enterprises in the constituency. Some of the councillors are supportive towards environmental movements but the majority lost interest in the subject once elected. Some maintain the view that Russia needs sacrifices while having economic problems and the environment cannot escape its share even if it becomes the victim of this development. A few councillors, however, try to give as much support to environmental groups as they can. They are, however, very much in a minority among councillors. Consequently environmental movements in Russia find themselves in very difficult circumstances. They face overwhelmingly unsupportive authorities both at local and national level. They are deprived of information, documents and data which could help in their arguments and are faced with a lot of hostility. Their declining public support helps them to avoid harassment from the authorities, which choose simply to ignore them due to their political weakness. When it comes to battling with the authorities, environmental movements in Russia are almost always the losers in this one-sided struggle.

9

A Comparison of Russian and Hungarian Environmental Movements

Introduction

In this chapter I will identify the major similarities and differences between environmental movements in Russia and Hungary. The comparison will emphasize the extent of diversity within the former Soviet bloc. I will discuss the differences deriving from the political context as well as the situation characterizing the movements. Russian and Hungarian environmental movements will be systematically compared in terms of their structure and goals, the characteristics of the participants and leaders, the extent of internal conflicts of the movements, their survival tactics and achievements, and the role of the media. Finally, I will examine how the different development of local and national authorities influences the features of environmental movements in the two cases.

The chapter thus has two major aims: as well as comparing Russian and Hungarian environmental movements I shall also discuss the extent of development of democracy because of its influence on the chances of success of grassroots organizations, such as environmental movements.

A Comparison of the Political Context

Before the Regime Change

Descriptions of the political context in Russia and Hungary often emphasize the similarities between them. In my view, however, the differences outnumber the similarities. It is not disputed that there was a period when the two countries' political development coincided in many ways, which was an intentional outcome of the leaderships' aims. The resemblance was strongest at the time when Stalin was in power in the Soviet Union and Mátyás Rákosi in Hungary. Between 1948 and the mid

1950s Hungary was shaped according to the Soviet pattern, as were most other socialist societies. The events of 1956, however, put a halt to this process and Hungary started to diverge from the "Soviet model" as far as circumstances allowed, as analyzed earlier. This allowed Hungary to introduce economic reforms. These were initiated originally as early as 1953 but were only taken seriously in the early and mid 1960s, with the official introduction in 1968 and were called the "New Economic Mechanism". Unfortunately the Soviet Union itself (after Khrushchev's removal from the political scene) did not manage to maintain the liberalization process and turned backwards politically speaking, especially in her domestic politics, while in Hungary— with some hiccoughs—the reform tendency remained unaffected on the whole and continued till the end of the regime.

Major political changes re-started in the Soviet Union only after the death of Brezhnev, although some relaxation in political terms had occurred in the period before his death. These changes stemmed from the recognition that the country's economy was stagnating and living standards were not improving. Examples of other socialist societies, such as Hungary, where the primacy of the Communist party was not questioned and the basic structure of the regime was untouched but where economic reforms and political relaxation had resulted in a better economic situation, higher living standards and less discontent than in the Soviet Union, encouraged the new party leader, Gorbachev, to replicate these changes. He initiated radical changes which led to a political upheaval, resulting in the so-called glasnost and later perestroika. The restructuring within the Soviet Parliament and local government aimed at increasing democratization within the Soviet Union under Gorbachev.

Gorbachev was too radical and too conservative at the same time. This provoked forces both on the right and the left. His conservatism was demonstrated in the fact that he dithered for too long over the introduction of a multi-party system in Russia. By the time he gained power, political demands were far more radical all over the socialist world and radical changes could not be restricted to reforms within the old regime. Failing to recognise this and insisting on keeping the leading and exclusive role of the Communist Party of the Soviet Union until the very end of his leadership, Gorbachev provoked the pro-democracy supporters, who as a result of the very glasnost which he initiated developed a recognition that a multi-party system is a necessary basis of democratic development.

Gorbachev's radical reforms, on the other hand, also alarmed the hard-liners who wanted to return to the political system as it was in Brezhnev's time. This led to the coup of 1991 followed by Gorbachev's lost popularity among the masses which brought his opponent, Boris Yeltsin, into power.

Developments in Hungary were very different in many respects and, as a result, Hungary did not go through such a stormy process as Russia. The reforms within the economy were followed by liberalization in the political sphere from the late 1960s onwards. The crises in the early and mid 1980s, when many reform supporters expressed worries that (a) the reform process was starting to slow down, and (b) that after Kádár's succession it could even be reversed, just as in the Soviet Union after Khrushchev, were at the centre of discussion in many social science publications. The development in the Soviet Union, the Andropov-Gorbachev political line, however, gave some hope to supporters of reforms in Hungary. Even when Kádár was replaced by Károly Grosz, who had the reputation of being a hardliner, the wider political circumstances could not keep him in power for long and he was swiftly replaced by a reform communist, Miklós Németh. This reform leadership within the party, and the government, peacefully led the country out of a one-party system.

In Hungary private property, including land ownership, had never been completely abolished and private business had gradually been encouraged since the 1960s with increased encouragement from the very early 1980s. State companies were also run by more competent management due to the New Economic Mechanism (Hare, 1977; Radice, 1981; Galasi and Sziráczki, 1985) and profit making by individual state enterprises became increasingly important (Soós, 1986). Hence the economic changes towards a market economy were much smoother than in Russia. In the political sphere the changes were also more gradual. Political parties started to develop at a very early stage of the transition (Bozóki, 1990), with rapidly established and recognizable political lines, unlike in Russia. The distinct political differences among the major political parties allowed the Hungarian electorate to choose with a fairly clear conviction by the first general elections in March 1990.

In Russia, however, both the economic and political changes followed a different pattern. The lack of gradual development towards a market economy combined with the political turmoil during the Gorbachev period resulted in serious economic decline. Living standards, which were already lower than in neighbouring European socialist societies, were falling and phenomena appeared in the life of Russian people which were completely new to them, such as high inflation, unemployment and privatization. These were features which had been known in socialist Hungary since the introduction of the New Economic Mechanism in 1968.

After the Regime Change

After the elections of 1990 Hungary embarked on a major reconstruction of the political system. The multi-party elections were successful in the sense that they brought in a government for four years with fairly clear political goals. The government was a coalition of three parties of the centre right and was nationalistic in emphasis and moderate in style. The opposition parties were also well organized and not much behind the victorious ones in popularity, as expressed in election results, and in their representation in parliament. Radical changes were introduced in the central administration as well as in local government. The first and most important intention, based on an all-party consensus, was to give the elected political bodies the leading role in national and local policy-making. Representatives were made accountable to their electorates. National and local elections were set for every four years to ensure a cyclical system of democratic elections both nationally and at local government level. Parliament became the most important body of political decision-making at the national level. The system of local administration was also radically reorganized. It was decentralized, delegating power to the lowest level of local government. Local councils, which were artificially united against the wishes of the local population during the state socialist period, were decentralized. As a result, the number of local councils has been doubled. More importantly, local authorities gained responsibility for their budgets, as well as policies, independent from each other. The previous strict hierarchical system has also been abolished.

The legal system became politically independent, based on the existing criminal and civil system. Solicitors had been acting on a private basis even prior to the democratic changes, but were not completely free politically. Judges and prosecutors were politically dependent, a system which changed from 1990. Parliament embarked on a continuous process of replacing the outdated socialist law by new legislation. This process will need considerable time to complete, especially as certain issues, such as social policy (pensions, health service, and so on) and the economy are more frequently discussed (and changed) than others, such as the environment, which has no priority for any government or major party in Hungary. It is only parliament, however, which can issue new laws in Hungary and only elected bodies, such as local authorities, which can issue legally binding regulations.

Overall, every attempt has been made to create a society where political structures and processes follow those developed in "older" democracies. These processes did not, of course, take place without major errors. The newly elected government in many ways echoed the Communist practice. Former loyal communist civil servants, for example, were replaced by new ones, but in many senior cases by those who were "politically loyal" to the new government. Similar changes occurred in the leadership of important institutions, like national banks. The most damaging attempt to establish political loyalty occurred in the media. The state television and radio presidents and a large number of journalists, who were known critics of the new government, were sacked. However, these measures backfired in two ways. Firstly, they drew attention to defects in the appointment process to such positions and secondly, they added to the rapidly growing unpopularity of the MDF-led government which contributed to their spectacular defeat in 1994. The problem of partisan civil servants and managers of nationally important institutions, media leaders and journalists still exists and has high priority in political and parliamentary debates. The MDF-led government lost popularity only six months after the national election in March 1990 to such an extent that it brought about an overwhelming victory for the opposition parties in many local governments in December 1990, which created a very interesting political situation.

In Russia, on the other hand, it was only the last years of the Gorbachev period which addressed and, to an extent, implemented changes which led to a more developed democratic system. After Yeltsin replaced Gorbachev as president of Russia, however, no attempts were made to continue the process of democratization, initiated by Gorbachev through reorganising the parliament, the constitution and the system of local government. Apart from allowing political parties to compete during elections (but not to compete on equal terms, for example by allocating them equal shares of television time or a fair amount of financial resources) several direct attempts have been made to centralize power in the hands of the president and non-elected leaders and no attempts have been made to place political power in the hands of elected bodies (*Heti Világgazdaság*, 1994a; 1994b). The electoral system does not lead to a proportionate representation of the elected parties in parliament, and parliament is not the only body which issues law. It is the president and mayors, who are sometimes not even elected but appointed by Yeltsin, who issue decree after decree (Tolz, 1994; Vida, 1995).

At local level it is not elected bodies which make policies, issue regulations and are in charge of the budget allocation, but the unelected executive part of the authorities. There were no attempts to replace civil servants and local government officials "inherited" from the Soviet period. Hence the majority of the Russian apparatchiks follow routines acquired in the old regime. As they are "in charge", in both political and executive terms, of the old apparatchik system, (originally developed in Tsarist Russia), it continues more or less untouched, as discussed previously.

The legal system in Russia is not independent either and it remains (as before) completely inadequate to fulfil its tasks. The lack of civil law (never developed in the Soviet period), the chaotic legislative system, in which contradictory laws are frequently issued (often by the president) without systematically replacing outdated ones, are just some examples characterizing the situation (Perepjolkin, 1997).

The decentralization process of the local administration, introduced under Gorbachev in 1990, has since been reversed (Boyce, 1993; Hanson, 1993). The frequent reorganization of both local and national administrations shows two clear tendencies: firstly a re-centralization, a shift of power back to the top level within the hierarchy and secondly, a shift back towards the hands of non-elected bodies. The media, which under Gorbachev experienced a refreshing openness, has become closely controlled and censored again. This is also true of the so-called independent papers, and there is no sign of any relaxation in this field.

To sum up, Hungary has been through a gradual process of reform concluding in a peaceful and fairly smooth transition into a multi-party system with a market economy. Political parties developed at an early stage during the regime change and the all-party consensus led to a gradual and steady development of a democratic structure. Hungarians have been more accustomed to "capitalist" phenomena, such as price rises and private property, which became familiar features over the decades after the New Economic Mechanism was introduced, but which are very new for Russians.

By contrast, political and economic changes took Russia by storm. The openings and democratization attempts under Gorbachev have largely been reversed. Many Russians, including political leaders, see the solution to the evolving difficulties, both in the economy and in the political sphere, in the concentration of power. A multi-party system has been introduced and different political parties could compete by the December 1993 general elections, but the electoral system does not ensure proper representation. Elected bodies on the whole do not play the leading role; instead it is the non-elected executives who govern the country at all levels of administration and they are often members of the "old" apparatus. Neither democracy nor a market system has been established in Russia, while social policy achievements, established in the Soviet period, are rapidly being eroded. While most people's living standards have been decreasing, social polarization has increased greatly. This is a subject of deep dissatisfaction among most Russians who still maintain a strong belief in social justice developed in the Soviet period. No serious attempts have been made to extend the process of democratization and, in fact, all the evidence suggests that the most important trend is just the opposite in Russia today. A concentration of power and a process of re-centralization is occurring under Yeltsin, a process which was re-enforced by his increased presidential power, voted for overwhelmingly by the 52 per cent of the population who did vote in the December 1993 elections (Sakwa, 1994).

A Comparison of Environmental Movements

The upsurge in concern over environmental problems which was emerging in the developed world in the early 1970s reached Hungary fairly soon after it occurred in the developed countries. It, however, avoided the Soviet Union which was more isolated than Hungary. Concern over the environment developed in Russia prior to the revolution as a result of nineteenth century German influence (Weiner, 1988) and this continued in the Soviet period, independent of western influence. This romantic notion regarding Russian nature characterized the development of environmentalism up to the late 1980s, when it started to combine with ideas arriving from the west.

Before the Regime Change

Social movements existed in both societies before the political change but in Hungary they were scarce, even if political liberalism was more advanced than in the Soviet Union. In Hungary the very few social movements concentrated on peace and environment related issues. It is interesting that organized movements were not allowed to develop by the ruling Communist leadership, considering that political liberalism was the greatest in the entire Soviet bloc. Critical voices were strong, and opposition existed among intellectuals and workers as well. Hungarian dissidents experienced much less repression than their counterparts in the Soviet Union, but independent grassroots organizations were repressed.

In the Soviet Union, in contrast, the political repression was greater but movement activities became stronger than in Hungary. Political opposition existed in both countries but in different forms. The Hungarian opposition was more focused on political issues. Lack of freedom in the press and publications and the prohibition of free assembly were the main issues among intellectuals, and pay and not enough say in management decisions was the main concern among workers' opposition. In the Soviet Union opposition mainly concentrated on nationalism and religious issues or else it focused on collective issues, not catered for by the "all embracing" state system (neighborhood movements, interest groups). There were other groups which were turned into opposition groups by the political leadership, including pop groups, art groups, environmental protection groups (to save Lake Baikal, Siberian rivers, Lake Aral, the forests in the north, for example) often led by famous writers concerned with the deterioration of the Russian environment.

After the Regime Change

Environmental movements started to mushroom in the two countries at the same period as a result of the parallel political changes. Gorbachev's glasnost and the relaxation of the last Communist leadership in Hungary were closely connected and resulted in an upsurge of social movements in both societies simultaneously. Environmental movements were in the vanguard of social movements in both societies, combining oppositional political sentiments with concerns over environmental problems. Many activists of this early period catapulted to political fame and later became leading politicians (in different parties of the right and of the left) having participated in environmental movements first.

In Hungary the process of differentiation between greens and politicians occurred at an early stage. A strict distinction between party politics and non-party political activities very soon followed the short period of transition. The sphere outside party politics, social movement participation, is therefore sharply distinguished by Hungarian movement activists, who consider themselves to be political activists in a non-partisan civil society.

In Russia this separation between political party activities and social movement activities has not developed to the same extent as in Hungary. The concept of civil society has not been discussed or emphasized either, in the same way as it has in Hungary. Hence in Russia party political activities are not looked upon as a completely different type of political engagement from participating in civil society actions. Russian movement activists find it easy to reconcil, being part of party politics (especially in opposition) with being active in social movements. This can be related to the fact that political parties in Russia are much more unstable than in Hungary and most of the opposition political parties are much weaker. This led to a situation in which different political forces, (political parties and social movements), all in opposition, wish to combine forces rather than remain divided, as in Hungary. In Hungary, on the other hand, after 1990 various political parties which were in clear political opposition with each other, have entered into government. Political parties do have a chance to reach power. Grassroots opposition in Hungary distinguishes itself as a political force which does not aim at political power but remains in permanent political opposition to any government.

Both Russian and Hungarian environmental movements, however, went through a stage when mass demonstrations and meetings were a frequently used method in expressing views and putting pressure on relevant authorities. This has changed in both countries for different reasons.

In Russia it not only disappeared from the arsenal of political weapons but has become a condemned method by everybody: the public, the movements and the authorities. In Hungary however, demonstrations have not been rejected by any means, but a certain "demonstration-fatigue" occurred after a while which led to growing difficulties in mobilizing the masses. This was, however, looked upon as a lamentable outcome by all, especially by senior Hungarian politicians in charge of the environment, who even blamed the movements for not trying harder to mobilize spectacular demonstrations.

In Russia there is a much stronger desire for collectivism than in Hungary. This is expressed in the case of environmental movements in the strong wish to unite in federal organisations. Russian environmental movement activists devote considerable time and energy building and maintaining federal organizations, which do not exist in Hungary. These federations help the individual movements to exchange information and provide support for each other. This desire to unite is, however, a clear sign of political weakness. Individual movements in Russia do not feel forceful enough to face authorities or enterprises in their fight and hope that united forces will provide stronger support.

In Hungary, on the other hand, no umbrella organization exists for environmental movements and no individual movement expressed the view that there was a need for it. There have been tentative attempts to organize such bodies but they have all failed almost at birth. I would argue that in Hungary a different culture developed historically from that in Russia. This is connected both to the more recent post-socialist period compared with the Soviet case and to the post-socialist period. Since the mid-1960s under Kádár there was a special approach to "socialist values" as discussed above, which prevented another 1956 occurring in Hungary. This led to a culture in Hungary which is much more "individualistic". It also created a situation in which hierarchically built organisations are not desired, unlike in Russia. There is a good information flow and frequent contacts among the different Hungarian environmental movements horizontally, and they also meet at yearly conferences, but the idea of a hierarchical structure is strongly rejected by the individual environmental movements in Hungary. Hierarchy is not favoured either as a principle or as a practice.

The Movement Participants in a Comparative Perspective

The participants in environmental movements in both countries are very similar. The movements attract many highly educated people, frequently with a degree in natural science. At the same time environmental movements accommodate a large number of activists from all segments of society with all levels of education in both cases. The difference, however, is that in Russia the movements at a higher level within the hierarchical system of environmental movements (federal or other umbrella organizations) concentrate the most highly educated members. Local movements are more likely to have a few well educated members and for the rest of the activists to have lower educational background. In Hungary, on the other hand, where the hierarchical system does not exist, the distribution of well educated people and lesser educated activists is fairly even among movements.

There is an important difference in the age distribution of activists between Hungarian and Russian environmental movements. In Hungary most core members are middle aged and there is a large proportion of young people participating in the movements because Hungarian environmental movements pay special attention to attracting young people. In Russia, however, there is no such conscious attempt and consequently young people are not represented in large proportions within environmental movements as a whole. The age division between national and local movements is influenced, again, by the hierarchical structure of Russian movements. Local environmental movements often attract the older generations, (retired people), while national organizations are more likely to have middle aged activists with the exception of Greenpeace Russia, in which the average age of the activists is much lower.

In both countries there was a high proportion of female participation in environmental movements at every level. It is Russia where women were more likely to become leaders in both national and local movements. In Hungary, however, the proportion of women among leaders was not always proportional to their ratio as activists within the movement where they are the majority of participants.

According to recent literature there is growing concern among social scientists about the rapid decrease of women's participation in political parties and governments in eastern Europe (Einhorn, 1993). The disappearance of the quota system, which existed during the socialist period, has radically decreased women's previous ratio (around 25 per cent) in central and eastern European parliaments. Environmental movements, however, provide good evidence that women have not given up political activism in eastern Europe altogether, though their under-representation in political parties remains an alarming trend even though women are active at grassroots level.

The structure of each movement in both countries follows the same pattern. Firstly, there is a core of very active members. In Hungary many more of them have a chance to be paid for their activities after a period of voluntary work than in Russia where only very few movements had the resources to afford this. The second circle is the group of activists who take part on an intermittent basis and do not have regular responsibilities. The third group of people in both countries is the circle of sympathizers. The widest circle is the non-active sympathizers. In Russia, in the national and local elections in 1990, for example, many candidates were nominated by environmental movements and many of them were elected, which is a evidence that a wide circle of sympathizers exists. In Hungary, it is also true that environmental movements expressed their support for individual candidates at local and national elections.

The Movement Leaders Compared

In both societies the movement leaders are always chosen from the core members which is understandable: they become leaders because they are active, willing to devote the necessary time and energy and harismatic enough to be accepted for leadership. Leaders of the environmental movements studied were without exception highly educated people, irrespective of whether the movement was national or local, in both countries. This is because expertise is always regarded as essential to leading an environmental movement in its struggle.

However, leaders differed in terms of age in the two countries. In Hungary they were always middle aged people, while in Russia in some cases they came from the older generation. This is a consequence of the age characteristics of many Russian local environmental movements.

Hierarchy and Bureaucracy in Connection with Environmental Movements Compared

There were two important respects in which Hungarian and Russian environmental movements differed substantially. These are the questions of hierarchy and bureaucracy. Even though Russian environmentalists frequently emphasized their dislike of any hierarchical system as a principle, when referring to their movements, paradoxically the leadership structure of the individual groups was very hierarchical, as is the whole system of movement organization in Russia.

In Hungary there were instances of collective leadership where responsibilities were divided and shared and decisions were made collectively. In Russia the idea of collective leadership was raised on a theoretical basis only but has not developed in practice. There were attempts to avoid a hierarchical structure, but these simply meant renaming different functions euphemistically. Collective leadership with shared responsibilities has not been successfully implemented in the Russian environmental movements' practice.

Bureaucracy was also looked upon as a negative phenomenon in Russia. This was often "translated" into not keeping proper records about movement activities. This resentment was not present in Hungary and record keeping was part of their everyday practice. Correspondence with the authorities, as well as records of activists, were kept most of the time on computer files. In Russia the lack of elementary infrastructure, such as computers or offices, also prevented the activists from keeping proper files. The only exception was Greenpeace Russia, which was better equipped than average.

The Movements' Objectives in Comparative Perspective

Hungarian and Russian movements often differ in the way they came into existence. Most Hungarian environmental movements were started because of a particular objective they wanted to achieve. People became united because they wanted to stop a bridge from being constructed near their area which would increase traffic, or to stop nuclear waste being stored near their village. Russian environmental movements more often lacked a concrete short term aim (or else the concrete goal was only a triggering event for people with a common interest).

In Russia many environmental movements started off with a general concern regarding the environment rather than a concrete objective. The general state of the environment, either in the neighborhood or at national level, was the most important initial reason for many activists to unite at grassroots level in several Russian cases. All these movements started in the late 1980s. Unfortunately the tradition of general public concern over green matters did not persist and the few remaining environmental movements did not manage to maintain public interest either. Today in Russia the level of general interest in environmental problems is diminishing.

Different Characteristics of the Movements in the two Societies

Firstly, environmental movements in Hungary often changed over time while in Russian they mostly remained more static. Many Hungarian movements were mobilizing forces initially against a concrete, environmentally harmful obstacle, and then started to widen their horizons, unlike Russian movements which, with one notable exception (Bitsa), have not altered over time in this respect.

Secondly, Hungarian environmental movements often took up problems at a later phase, which lay outside their original scope, but became important in the course of their concrete struggle. This, after a while, became a conscious survival strategy for many of them. It helped the activists to maintain and continue with the movement even when they had attained their immediate task which would have meant the end of the movement or when they had failed.

Thirdly, the "Hungarian pattern" of change over time did not occur in the Russian cases. Instead those Russian movements which had general environmental concerns from the beginning continued as such and those Russian movements which started off with concrete aims did not transform into general green movements. As the process of openness was gradually reversed in Russia the movements' situation also became more difficult and the prospects of achieving their objectives (whether concrete or general) became much bleaker. "Russian" survival techniques therefore had to be less subtle and were frequently unsuccessful.

Fourthly, there was another important Hungarian specificity. This was the special attention devoted to educating (and hopefully recruiting) the population at large but especially young people. This phenomenon, which is stressed by Eyerman and Jamison (1991) based on their Scandinavian experiences, will be discussed later.

Fifthly, the process of qualitative change from one stage into another was also different in the two societies. The most important change was the denunciation of demonstrations as a method by Russian environmentalists, as discussed earlier, which did not happen in Hungary. Finally, it is worth pointing out that in both countries environmental movements also became important as an activity in itself for many activists, a sort of social club.

The Movements' Internal Conflicts Compared

There are conflicts within and beyond the movements in both countries. Internal conflicts were mostly based on personality clashes. Movement leaders are often people with a strong character and determined views and this often leads to strong debates. Whether compromises were achieved or not depended on the leaders' personalities as well as the members' willingness to conform or compromise. There was no difference between Russian and Hungarian characteristics in this respect. There were examples of charismatic and patient leaders in full agreement with the members and instances of irreversible conflicts leading to splits in both countries. Environmental movements which survived could lose participants for all sorts of reasons, not only because of internal conflicts.

One source of internal conflict could have been the question over party politics among the activists. As discussed earlier, political parties formed parallel to the movements which could have resulted in serious internal divisions. The environmental activists, just as the rest of the population, chose a political party when they were formed in 1989 in Hungary and 1991 in Russia. In a western context environmentalists are often associated with people who are more likely to have left wing political views. Russia and Hungary differ from the western situation, but not from each other, in this respect. Both moderate right and left wing views are present in both countries within one environmental movement. This in Russia mostly means people who either support reforms or the old regime.

In Hungary people are divided politically according to whether they favour a right wing, nationalistic political party, liberal views or the socialists. Environmental movement members have developed their political preferences over recent years and differ from each other. This, however, has not led to conflicts or splits within the movement itself.

Movement activists in both countries coexist with activists with diverse political views, supporting different moderate political parties. This was a principle which they all emphasized and a practice which was followed.

The only exception was towards polarized political views. And this is the point where the two countries inevitably differed. The political context was different. In Hungary polarized political views are scarce since only a small minority of the population voted for the extreme parties either on the right or on the left. Among the environmentalists it did not even appear as a problem due to the absence of extremist views among people associated with environmental movements. There are two, very weak and unpopular, green parties which are accused of being eco-communist and eco-fascist respectively and these absorb the minority who have such views (and gain extremely little attention or popularity among the public, to judge by their very meagre election results).

In Russia, however, fascist political tendencies are more popular than in Hungary. The once widespread popularity of Zhirinovsky's party (1993 general elections) is strong evidence to support this view. Environmentalists also faced the problem when, as a result of the rapid growth of eco-fascist ideas, the existing greens split. This became apparent when the eco-fascists formed a united party and rapidly gained popularity in several regions, such as St Petersburg and Chelyabinsk. Experiencing the speedy growth of eco-fascist ideas, Russian environmental movements faced the question of whether they wished to accommodate such views. All the movements we came across strongly resented eco-fascist ideas and individuals, and expressed their wish not to cooperate with them. In this respect there was no difference between the two countries. There was, however, an enormous difference in the rate of success between Hungarian and Russian environmental movements.

The Rate of Success of Environmental Movements Compared

Success and survival can be related. There are exceptions but generally movements are more likely to survive if they are successful. However, a comparison of the two cases here can provide a very interesting contrast.

Environmental movements in Hungary are fairly successful if we measure success by winning concrete fights and even more successful if measured by their growing fame and success in raising environmental consciousness. Hungarian environmental movements are also often successful in the sense that they have political influence locally.

In Russia, however, the situation is very different. The very low rate of success the movements manage to achieve is due to a combination of circumstances, which mostly lie beyond the actual movement. Facing serious problems in the economy was a good excuse for the "official" propaganda pursuing a strong (and penetrating) argument that environmentalism cannot be the major concern for the time being. This led to a situation where green issues are increasingly viewed as less important matters than economic problems. The short spell in the late 1980s when this was different seems to be over in Russia today. Environmentalism has become a secondary problem compared with others in the public's mind in Russia. As a result economic lobbies have become very strong again. And as authorities are dominated by non-elected bodies they do not react to organized pressure groups and their priority is also old-fashioned: the apparatchiks always strongly supported the economy lobby, as will be discussed below.

It is not surprising therefore to find a very low success rate among environmental movements in Russia. After the above mentioned short period when the public became more sensitive on green issues, and also wished to and could put pressure on the local and national authorities, and their efforts resulted in some concrete achievements, the movements experienced a rapid decline in success. This lack of achievements, both in concrete and indirect terms (popularity), became the most important trend among Russian environmental movements for no fault of their own.

I cannot argue that Hungarians were any more experienced in their methods, or that they chose more easily obtainable aims. Nor are Russian demands any less rational than Hungarian ones. The only difference leading to such a low rate of success in Russia is in the political context. In fact, Russian environmentalists work under so much more difficult circumstances and are facing so many elementary problems, including the lack of resources, that their heroic efforts deserve much better results. But instead, unfortunately, they face growing hostility from the apparatchiks, isolation by the media and consequently from the public. The little hope of success also led environmental movements to become less popular: they are looked upon as "Don Quixote" characters, with noble but utopian ideas at best. The media also contribute to this view in a negative sense.

The Role of the Media in a Comparative Perspective

The support of the media or the lack of it is crucial in the development of public opinion which includes opinion concerning environmentalism. The situation in the two countries could not be more different. Although in Hungary the media went through a period when part of it was under strong governmental influence (1990-1994), even then they were not completely controlled or censored. Environmentalism was in any case one of those "lucky" subjects which did not irritate the government of the time and which therefore was not the subject of censorship once a multi-party system came into existence in Hungary. Environmentalism is, in fact, a very frequent subject in the Hungarian press, television and radio. Even though none of the political parties emphasizes green issues, equally, none is against them, hence no political pressure prevents journalists from writing about the subject. Environmentalism has become one of those "politically safe" subjects which can fill up air-time or newspaper columns without any risk. Consequently it appears in the media on a very regular and frequent basis (and has done even when the media were under political fire from the MDF).

The media, on the other hand, is consciously and systematically used by the Hungarian environmental movements themselves to build their own reputation and to gain publicity. This, in turn, results in most movements becoming known all over the country irrespective of how narrow or widespread their objectives are. Even relatively small environmental movements, organized in a previously little known, tiny village become sometimes widely reported about and gain publicity, just as major national movements (for example the case of Ófalu (Juhász, Vári and Tölgyesi, 1993)) The reports mostly portray the movements' efforts sympathetically which contributes to their reputation and in turn to their success. The contact with journalists is continuous: the movements' activists always felt that the journalists were easily approachable and ready to report about them. Because environmentalism as a subject is favoured in the media it is easy for the movements to put an issue which is important for them on the journalists' agenda. Conflicts between the media and the movements in Hungary only occasionally occurred.

The Russian case is fundamentally different. In Russia today the biggest damage is caused by "omission": by the lack of reporting about the activities of environmental subjects generally and the movements in particular. This is the result of strong political pressure.

At the height of glasnost the media in Russia experienced an unprecedented chance to report anything of journalistic interest. The political taboos developed in the Soviet era broke down and there was a refreshing upsurge of interesting, informative and uncensored articles in the Russian press. As a result of the lack of political control, environmentalism has become a frequently approached subject resulting in ever growing public interest in the state of the environment. Environmental movements also felt well appreciated by the press's attention to their activities and were satisfied with the reports about them. The public became better informed about the movements, hence more and more people joined them. As a result the public pressure grew and some modest successes were achieved. It was a good start.

After Yeltsin came into power the "honeymoon" for the media was over. This happened on several fronts. Political freedom has been gradually eroded and this has been combined with the introduction of commercial aspects in the press. Previously subsidised papers ceased to be supported financially and many of them turned out to be unviable in the new market conditions. Cheap tabloids took their place. Other, more serious, papers were financed by state-owned bodies which, in turn, re-introduced political control. This was also the case with major state-owned television and radio stations. In the renewed anti-democratic media situation opposition voices, such as those of the environmental movements, were not welcomed. Reporting about them also became risky and journalists avoided the subject.

Because local papers were considered less dangerous by the Russian authorities they are the only media forums which are likely to report about (local) environmental events and movement activities. Their circulation, however, is small and only reaches those living in the neighborhood. Local papers cannot have the same mobilizing effect as the national media, hence the authorities' "tolerance" towards them. Public opinion on the whole has not been altered by local news which can only deal with isolated cases. This is precisely why local papers are not considered a political danger and are not blocked from reporting about the movements. This is a good example of how "democracy" and "freedom of speech" operate in Russia since Yeltsin came into power in December 1993.

This overall process, however, contributed greatly to the decreasing public interest in environmental movements in Russia, as well as the lack of increase in environmental consciousness. Movement members themselves often felt less inclined to stay in the movements as public opinion changed. Their declining respect and prestige, the "Don Quixote" effect, resulted in a rapid decrease in the number of movement participants and sympathizers. As prestige, the number of activists and supporters decreased so did the potential power of public pressure: it became very weak. Hence the ever declining rate of achievements, discussed above. The only reason environmental movements and activists are not targeted by the authorities actively as "public enemies" is because they are considered too weak to worry about as oppositional forces. The media-vacuum around the movements and activists creates the desired effect as far the authorities are concerned.

Had they been in a stronger position, in the anti-democratic political atmosphere which is increasingly developing in Russia today, they would most likely be attacked more actively and openly by the local and national authorities using their power via the media (smear campaigns) and even the police. The non-elected apparatchiks, according to their open testimonies in the interviews, are ready to start ridiculing environmentalists any time, if they sense any potential political danger in them, and would conduct dirty tricks campaigns or resort to even stronger measures. As it stands at present, environmental movements are too weak to be considered as potential political rivals. Rather, they are left alone and the media is instructed simply to neglect them.

The Relationship Between the Authorities and the Environmental Movements in the two Societies Compared

The Attitude of Elected Politicians

Environmentalism is a fairly neglected question by both governments. This is because no major political parties take the subject of environmentalism seriously enough to keep it high on the political agenda in either country. It does not appear as a central issue, even in party manifestos, either in Russia or in Hungary.

Green parties are not strong enough to send a single representative into parliament, again in either country. They are also divided politically into "red greens" or "water melon greens" (as the saying goes: outside green, inside red) and eco-fascists. Environmental movements follow a more moderate political route than green parties in either country. In Russia, however, the contacts between the movements and the left-wing oriented green party is stronger. In Hungary there is no connection between the green parties and environmental movements.

In Hungary there is a major difference between the attitude of elected politicians and that of the executives both nationally and locally. Hungarian elected representatives are accountable to their constituencies and pay a lot of attention to their re-electability. They are also in charge of policy-making and budgeting. Hungarian politicians see environmental movements as the best potential allies in gaining support and attention for the cause of environmentalism. By feeling the weakness of the issue within the political spectrum and needing public support Hungarian politicians see a potential in the environmental movements who can mobilize political sympathy and change public opinion. This is why Hungarian politicians expressed such a strong wish to cooperate with the movements and are also willing to support them financially. The hope that the movements would provide support in return for lobbying together (or at least for the same subject) pulls politicians and movement activists together.

Hungarian movement activists, however, look at the problem from a different point of view. They are sometimes hostile towards politicians and see them as party political people first and environmentalists second. But it is also recognised by movement activists that cooperation is more worthwhile than antagonism and therefore they are often ready to compromise.

Meanwhile, Hungarian politicians are disappointed that the movements generally do not attract the kind of public support, manifested in mass demonstrations, they had before (at the turn of the decade) and that Hungarian environmental movements do not form national umbrella organisations (like Russians) which could be presented to the government as stronger evidence of public support. Nevertheless, politicians try to cooperate with the best known movements as best they can. Thus both movements and politicians in Hungary provide each other with information, and are ready to exchange ideas and collaborate.

In Russia, on the other hand, politicians are much more divided on this question. At the national level there is not much positive interest in environmental movements. In fact, the relationship should be described as hostile and antagonistic.

At local level, however, there is a clear recognition on the side of some elected representatives of a possible role for environmental movements if properly supported. Some councillors try to help them as best they can in their own meagre circumstances. However, as local representatives themselves do not have much power their support does not lead environmental movements very far. These councillors are ready to share the limited information they can get hold of and express support on the whole.

There are exceptions, however. Some councillors, elected because of the environmental priorities they had expressed during the election campaign, changed their priorities once elected. On some occasions they accepted ideas propagated by the economic lobby. When they were in charge of analyzing concrete cases put to them by an environmental movement and these councillors were asked to prepare recommendations for the apparatchiks (as ironically is the reverse way of decision-making in Russia), the economic importance took priority over some councillors' views in the environmental committees. They did not accept the environmentalists' demand that industrial enterprises should not be located in highly populated areas and forwarded to the apparatchiks their recommendations of rejection. This kind of change of heart was frequently mentioned by environmental movements as well: the environmentalists also explained that this is a recurrent event which often happened to former environmentalists, who had become councillors or MPs.

Thus the two countries show a very different pattern in relation to the elected representatives and environmental movements which has its origin in the different ways of treatment and role of elected representatives in the overall political system.

The Attitude of Non-Elected Officials (Apparatchiks)

Now I turn to the comparison of the non-elected officials or, in the Russian term, apparatchiks. Non-elected officials in both countries have a more negative attitude towards the movements. The most important difference here, however, is that in Hungary they have a lot less power than elected representatives. In Russia apparatchiks play by far the dominating role, and increasingly so.

In Hungary an official can have a highly negative personal view of an environmental movement, but that cannot stop him/her from having to fulfil the politicians' demand to cooperate with those movements. Some ministerial civil servants can and do, for example, dislike or dismiss environmental movements and did not hide this in the interviews. They are also clear, however, that they are obliged to heed the wishes of the minister of the environment. This is similar at local level where the district environmental committee leaders or even the mayor, who is also elected as head of the local authority, is in charge of policy-making.

Often, however, there was a good relationship between the movements and the executive part of the local authority (in Budapest City Council, for example). It is perhaps the civil servants at the regional level who are the least likely to be keen on the movements' activities, and some provincial city council officials. Some of them are professionals who see the environmentalists as amateurs, people with less knowledge in the subject. Being in the regional or provincial offices, a little further away from direct political instructions than civil servants in ministries or local councillors, these officials tend to follow political guidelines only as much as they have to and at the same time they look on environmental activists as "impossible busy-bodies".

In Russia as established, non-executives keep all the power in their hands, (and this is increasingly so at every level). As they are not elected people they have no interest in keeping up a good image and consequently they do not keep good contacts with environmental movements and have a very low opinion of them. No potential cooperation could be envisaged as the officials, according to their testimonies, do not see any point in working together with the social movements. In fact not only social movement activists but even elected representatives and political party activists are all seen as one category in the eyes of Russian apparatchiks: they are all irritations, they are all viewed as part of the very uncomfortable outcome of the new system caused by a very uncomfortable upheaval. This new "democracy", argue the apparatchiks, only forces them to live with this group of "trouble-makers", who have a different shape and form but are basically all the same: they are all in opposition (a very dirty word in the apparatchiks' vocabulary). These "trouble-makers", including elected ones, only prevent Russian officials from creating "order". These people should be avoided and excluded from power at any price, is the apparatchiks view on the matter. They are at best nothing but misguided political intruders and think that in "democratic" Russia the system would or should be different then before (at any time in history), argue the apparatchiks.

These so-called "troublemakers" are dealt with one by one, and sadly, fairly successfully by the Russian apparatus at all levels very consistently. Giving up power and position, getting into complicated debates over priorities or offering any sort of cooperation or political compromise is not what the apparatchiks were used to in Russia during or prior to the Soviet period. Consequently the apparatchiks do not intend to show any sign of change. They are convinced that "order" will only be re-established in Russia if it is again dictated by the apparatus as in the Soviet era. The so-called "democratic" system existing since Yeltsin came to power reinforced the "old guard" of apparatchiks in their position to stay on unopposed, unelected and untouched. As a result they pursue their functions in a continuous manner.

As there is no attempt to create a politically neutral civil service whose job is to implement policies devised by elected and accountable politicians in Russia, and as independent political forces such as pressure groups, opposition party activists and elected representatives will never be able to put pressure on unelected officials, the development of any "proper" democratic process does not look very likely in the Russian case at present.

The way apparatchiks relate to opposition, which includes environmental movements, is not new. They are only following all the old routines they used to pursue before. As there is no pressure radically and fundamentally to reform this system (as is happening gradually but much more forcefully, hence successfully, in Hungary), nothing about it will change in Russia in the foreseeable future.

Thus the key to the problems of Russian environmental movements lies outside the environmentalist movements. The lack of proper democratic institutions inevitably creates a situation in which independent pressure groups in opposition are doomed to fail. This is the case of Russian environmental movements.

Conclusion

To sum up this chapter, I would argue that the fact that Russian environmental activists feel very pessimistic and disillusioned in many ways should not be of any surprise to us. The complete lack of support or even willingness to communicate by the authorities naturally leads to those negative feelings on the side of activists. By withholding public documents, not allowing the movements to obtain the information necessary to prepare a fair argument, the lack of help in funding, negative attitudes and a hostile atmosphere leading to a series of failures or even open fights, make environmental movements work in very difficult circumstances.

Even though, as explained above, Hungarian officials are often fairly unfriendly themselves, Hungarian environmental movements are much more positive when describing their chances vis-a-vis the authorities. There is, again, very good reason for that. The apparently famous Hungarian pessimism does not apply here, there is no grounds for it. The regular exchange of information, the funding and cooperation combined with a fairly friendly atmosphere with the politicians and a lot of support even from some officials, make Hungarian environmentalists feel wanted and needed, and their efforts worthwhile. Added to this they have achieved numerous successes which makes them feel the potential and will for further battles. The public, thanks to media support, is behind them, and their activities and achievements are widely publicised. Hence many of them widened their range of interest and took upon new tasks, such as education in a wider sense. It is clear to most environmentalists that they had to compromise with the authorities to be able to cooperate with them, but it is seen as the right way to pursue matters. Politicians are viewed with a certain degree of suspicion, as people who are often nothing more than careerists, but that does not exclude a willingness to collaborate with them for the sake of the movement's aims.

Finally, I return to our opening statement. I have argued at the beginning of this chapter that it is the political context which is the most important basis within a society. The two cases of the former Soviet bloc have shown us two very different examples of democratic development since the regime change. The Hungarian case can encourage optimism about the future but the Russian case is at present rather gloomy. Tolerating political opposition is the basis of democracy. In Hungary the system has been established and is working. This allows not only opposition political parties to operate but non-party political opposition as well. The case of Hungarian environmental movements provides a good example of successful opposition. In Russia, however, democratic institutions have not been established. The period since Yeltsin came to power shows a process of recentralization and even less development in democratic processes than in Gorbachev's time. Not only the opposition but even elected representatives are politically ignored in Russia and are powerless. Not surprisingly, non-party political opposition also is having a difficult time as the case of Russian environmental movements showed. I now turn to an examination of various theoretical arguments in connection with eastern Europe.

10

The Relevance of Western Theories in Eastern Europe

Introduction

In this chapter I will confront various theoretical approaches with my empirical findings. I shall argue that the theoretical works based on American and western European experiences inevitably leave unanswered questions when measured against eastern European cases but their usefulness far outweighs their shortcomings. Civil society theory is the only body of writing to have been based on eastern European experience as well as on western, and it will be argued that it too has some value in understanding the contextual conditions under which social movements develop.

The Relevance of Civil Society Theory and the Concept of Public Sphere in Eastern Europe

Let me deal with civil society theory first. The concept, as discussed earlier, has a centuries old history and has been applied in many different types of society from eighteenth century north America and Europe to present day eastern Europe.

There is a close analogy between the period when civil society became a relevant concept in western democracies, at the time when financial and commercial capitalism emerged and the elements of a new social order were taking shape in western and southern Europe, and the situation in eastern European societies in transition today. Then, just as now in eastern Europe, two parallel processes occurred at the same time: the development of capitalism in the economic sphere and the restructuring of the political sphere. There are several similarities here: firstly, the changing character of the ruling authorities, leading to greater freedom for the individual and secondly, that the public could challenge the state administration, and thirdly, that radical social changes accompanied this development.

However, we also have to be aware of the fact that state socialist societies were not feudalistic ones at the time of the regime change. When eastern European societies became socialist there were still strong elements of feudalism in all of them but the socialist regime undoubtedly brought about a tremendous amount of modernization as a consequence of which contemporary state socialist societies developed a curious mixture of a modern European state and an overcentralized, over-controlled society. And, as we argued previously, civil society (in my sense) was also present in state socialist societies, even if it existed within constraints. Freedom of assembly, association of any kind of organisation and freedom of the press were certainly not guaranteed. However, there were parliaments, a legal system and political parties, media and public opinion even if all of them were under a large degree of political control.

The realm of free and independent political protest existed under socialism, although it was curtailed. There were many associations which existed legally but some of them were in opposition. Other groups existed illegally. The press, which played such an important role in the development of western civil societies, was censored but censorship cannot be perfect. Both in the Soviet Union and in Hungary articles appeared which later caused serious headaches for the editors. The process of self-imposed censorship was much more successful in the Soviet Union than in liberal Hungary but no-one can deny the appearance of protest literature within the Soviet press, including the works of Solzhenytsin and many others. Political criticism also existed, again more openly in Hungary than in the Soviet case. Apart from that of course there existed an illegal samizdat literature which not only reached those in opposition but even the inner circles of the political leadership, which was its main target. Hence, as an opposition force, it certainly fulfilled an important political function within the limited sphere of civil society under socialism.

The question of freedom of association is difficult in the sense that, as I mentioned, many groups existed legally and shifted into opposition, or became perceived as part of the opposition. This was especially true for the Soviet Union, precisely because the political tolerance level was much lower than in Hungary. Groups or associations which would have been perceived as apolitical in a western democracy, or even Hungary, such as pop groups or art groups, were treated as subversive in the Soviet political context. In contrast, organisations which were perceived as bastions of the regime, such as the Young Communists' League (KISz) in Hungary, produced publications containing articles with strong criticism of the regime (Medvetánc).[1] As the line was fairly blurred one can only safely state that civil society, outside the realm of the similarly blurred state control, existed in both societies, as everywhere else in state socialist countries. Most of civil society activities were informally finding their ways within the unclearly drawn and constantly changing limits of political tolerance.

In Russia, even before the regime changed, the level of political tolerance was rapidly increasing, and people were ready and eager to take part in political protest activities. Civil society grew to an unprecedented extent under Gorbachev's policy of glasnost, with the formation of 60,000 "neformaly" (informal groups, as they were named in Russian) (Berezhovsky, 1990; Yanitsky, 1993b). These groups did not arise from nowhere, but had their roots in the pre-Gorbachev period, and their sudden upsurge was only due to the changing political atmosphere, which ceased to constrain them. Similarly the gradually growing freedom within the press had its roots in the Khrushchevian past waiting for the new impulse to reappear. In Hungary there was a more gradual development of reforms but in the late 1980s there was a sudden upsurge of political interest resulting in the mushrooming of social movements including environmental movements.

However, it needed a complete regime change to fulfil Habermas's definition. In guaranteeing the effectiveness of a public sphere in the political realm there are two important aspects to be fulfilled. Firstly, basic rights, such as a free press, freedom of assembly and association, and freedom of speech and opinion have to be guaranteed by the state (Habermas, 1992). And secondly, the state has to oblige organizations to fulfil their task and to structure their internal order accordingly to guarantee these basic rights. Thus the public sphere, which is part of civil society in Habermas's definition, can only be achieved by state guarantees. The question for us will be whether this has been achieved in Russia and Hungary.

After 1989 in Hungary the party political system stabilized and the institutions to ensure democratic development were established, as we demonstrated in previous chapters. Civil society also continued to develop. The period of transition from state socialism to liberal democracy encouraged civil society to spring up, as it did throughout the region. Civil society, like the whole system of democratic institutions, also found its function within the new regime. It carved out its role very clearly, separating itself from party politics, and embarked on an important protest role, while maintaining continuous interaction with both political parties and administrative authorities. Environmental movements became an important part of this process. They attracted movement members and a large number of sympathizers. They took on every level of the state apparatus and utilized the media successfully. The result is shown in their concrete achievements and in their high reputation among the public and the politicians, as has been demonstrated in this study. Habermas's criteria have been almost completely fulfilled in Hungary. The state does guarantee citizen participation in the public sphere: freedom of assembly and association exist, and freedom of speech and opinion is present. With some initial hiccoughs the media has also become free. Most importantly, state organizations are obliged to fulfil their task and to structure themselves internally to guarantee these basic rights.

The Russian case is different. Soon after the revolutionary period of Gorbachev when a gradual process of democratization started in Russia, as our evidence demonstrates, a recentralization process occurred. The process of democratization was reversed. At present in Russia political institutions do not fulfil the requirements of a democratic regime and no attempts are being made to change this. The trend in Russia today is backward compared with the glasnost period. The environmental movements analyzed here show well how impossible it is to achieve any success in an undemocratic regime. The lack of development of democratic institutions seriously hinders civil society: it can exist but it cannot be effective. The state does not guarantee citizen participation in the public sphere, and the media are not completely free. Freedom of assembly and association exist, as does freedom of speech and opinion, and in this sense the changes are fundamental, but state organizations do not fulfil their obligations in guaranteeing that these basic rights work effectively. Russian democracy therefore fails on Habermas's criteria.

My first conclusion is therefore that civil society has to be defined clearly, in my understanding, as separate from state and political parties. It is fundamental, however, not to treat civil society in *isolation* from state and political parties since it is the *interaction* between them that shapes civil society. I therefore consider that those writers of civil society who regard it as the exclusive key to understanding social movements and/or the state of democratic development in eastern Europe, or anywhere else, are mistaken. This leads to my second conclusion, that civil society can fully develop only in a democratic regime. This is in accordance with Habermas's (1992) emphasis on the primary role of the state on the public sphere. However, I find it useful, for analytical purposes, in modern society to separate the state and political parties from the sphere of extra-parliamentary political civil action. Finally, the concept of civil society should not be treated ahistorically: as Habermas (1992) also demonstrated in his account of the transformation of civil society and the public sphere in several western democracies, it is also important to realize that the different experiences in eastern European societies, as elaborated in the first two chapters, are the basis of the diverging realities of democratic development in general, and of civil societies, including environmental movement patterns, in particular.

As civil society is concerned with the context-setting and power-challenging aspects of citizen action, democracy, state and polity and the interaction between them, once clearly defined, the civil society concept is useful in demarcating a category of power-challenging phenomena and the conditions under which they arise. It does not, however, seek to answer how and why collective actions occur and function. This is the central concern of social movement theories.

The Relevance of Social Movement Theories

I now turn to the social movement theories—from the collective behavior approach, resource mobilization theory, new social movement school, political opportunity structure theory to the cognitive approach, discussed in chapter 4. Although social movement theories started a lot more recently than civil society theory, and consequently do not have to bridge a gap in time, since none of them have dealt with eastern Europe, their application in this context is innovative.

The Relevance of the Collective Behavior Approach

The most important feature of one of the social movement theories, called the collective behavior approach, is that it analyzes social movements firstly, as part of the very wide category of collective action and secondly, as social action. Both these aspects are relevant and important in any society today, as there is little agreement among the different authors as to what should constitute a social movement and it is important to realize that collective action is embedded in society. But, by lumping together all sorts of collective actions from crowd through panic to revolution, and embracing several forms of social action which have little in common in our view, this theory does not bring us closer to an understanding of social movements either in eastern Europe or in any other society (Smelser, 1962; Killian and Turner, 1972).

The collective behavior approach also looks at the different forms of collective action as a hierarchy in which crowd gatherings can develop into social movements (Smelser, 1962). This is an unacceptable interpretation for us. We do not see the early demonstrations, petitions and crowd gatherings in eastern Europe at the time of the regime change as predecessors of later social movements. In fact, social movements developed either earlier than or simultaneously with the peak of mass movements on the streets of Hungary and Russia and most demonstrations were organized by these social movements. Demonstrations, petitions and other forms of protest action are the tools of social movements rather than their embryonic forms, as collective behavior theory argues. On the other hand, the collective behavior approach's view of social processes as resulting from value transformation, as argued by Smelser (1962), is useful in understanding what is happening in eastern Europe today and why social movements occur. But the theory's functionalist approach, which treats collective behavior as abnormal, as a reaction to social breakdown caused by rapid social changes, is inappropriate in my opinion in explaining events in eastern Europe (or any society). Social movements did not occur because of social breakdown. They occurred because of a major opening in society, the ending of the political control which kept collective action at bay in the state socialist system. Collective action is not the result of anomie (Killian and Turner, 1972) but the result of the creation of democratic institutions which allow people to form political groups if they wish to express their protest against existing political routines. At the time when some representatives of the collective action were concerned with the analysis of the growing fascism in some European societies the conclusion of social breakdown was understandable. The situation in the United States, where it was later applied, and in eastern Europe today is, however, different.

To sum up the relevance of the collective behavior approach for our cases, we would argue that generally speaking this theory is not helpful in understanding social trends which are moving in the direction away from a dictatorship. In eastern Europe where, despite the Russian problems, the underlying trend is a move from an over-controlled political system towards a more democratic one, the "breakdown" theory of the collective behavior approach does not apply (Killian and Turner, 1972). In addition to that, neither the psychological nor the functionalist approach helps us to understand why social movements occur and how they operate. The attention to value changes (Smelser, 1962), leadership styles and success (Killian and Turner, 1972), however, turned out to be useful tools in our investigation and most importantly the fact that social movement activities are viewed within their social context. We now consider the relevance of the resource mobilization theory.

The Relevance of the Resource Mobilization (RM) Approach

The resource mobilization approach arose as a challenge to the collective behavior theory. The starting point of this new approach was therefore not the individual but the organization. The most important aspect of this perspective is how effective participants are and what resources they need to achieve their goals with the movement. The corner stone of the resource mobilization approach is seeing the rationality of human actors who are assumed to calculate costs and benefits when participating in social actions and are accompanied by the "free-riders" who let others take a risk and only hope to reap benefit from the movements' achievements. Thus the focus is on the instrumentality of movement strategy formation and on how movement organizations went about trying to achieve their goals. It is concerned with the collective-organizational level of analysis of movement organizations and organizational strategies by looking at mechanisms and incentives and tries to uncover the underlying rationality through a meso-level approach.

The resource mobilization approach's argument, that collective action is not abnormal, is a very important argument. We also see social movements as normal responses to the challenges of society in eastern Europe (and elsewhere) because it is not social "strain" which leads people to join social movements. However, we find it difficult to agree on the other side of the argument that it is rationality, based on cost-benefit analysis, which leads people to take part in collective action or decide whether to become free-riders. It was true in the cases of the "not in my backyard" type of movements, that participation was initially strongly motivated by self-interest but movements of these kinds were in the minority in both countries. And more importantly, many of the movements which started off as "not in my backyard" groups often changed their character and became environmental movements supporting a wide range of long-term green issues; thus rational calculative behavior was not characteristic. When it comes to costing the benefits, in fact, there were many more losers among participants than gainers. Many core activists in Hungary and Russia lost their jobs for their political views or by spending too much time and energy on the movement. Some in Hungary later managed to gain some financial support from the movement but it never came near to their previous income. The one very common motive among activists, in fact, was a utopian desire to achieve a better environment for all, a highly "irrational" and very long term way of thinking. Resource mobilization theory therefore was not proven relevant in this respect.

Another argument of RM theory is that organization benefits movements in achieving their goals. As was shown, Russian environmental movements are much more structured, and better organized than Hungarian movements where there is no hierarchical organisation at all. Yet Hungarian movements are considerably more successful than Russian ones.

The question of success was also crucial within RM theory. However, our understanding of success is slightly different from that of resource mobilization theorists'. We too incorporated among the many aspects of success achieving concrete goals obtaining changes in policies, and maintaining the group itself, as originally argued by RM theorists. However, our interpretation of success also emphasized the distribution of knowledge in the wider community which has never been considered by RM theorists. Success, nevertheless, was a central issue in our investigation, as well as that of the RM approach, and we found very useful the initial impetus of their analysis in drawing attention to it. They, however, failed to explain why success has been achieved in one society on a much larger scale than in another. None of the reasons the RM approach has advanced helped us to get closer to the answer to this most crucial question. This is because, in fact, the answer did not lie in the difference between the amount of resources used by Hungarian and Russian activists, especially not when we consider that their intellectual resources were, in fact, very similar. The differences in financial resources are the consequences of the lack of support, not the explanatory reasons in explaining the differences between Hungary and Russia. Thus material resources are of secondary importance compared with the fact that the entire societal context in which these movements operate is different.

The answer therefore cannot be found in a theory which views social movements as rational organizations, operating in isolation, and which explains everything in terms of features of the movement itself. Some aspects of the resource mobilization approach were useful when analysing the different movements but the meso-level of interpretation did not bring us closer to the real explanation which lies in the differences between the two societies. The theory which looks at movements not as rational mobilizers of resources to achieve certain ends but as transforming agents of political life offered for us a more useful analysis. This was the new social movement theory.

The Relevance of the New Social Movement (NSM) Theory

The social movements perspectives we have looked at so far have been developed in America. The new social movement theory was conceived in Europe, even if this "Europe" was, in fact half of Europe, only western Europe. NSM is the product of many trends in recent decades some of which are similar to tendencies in eastern Europe. There are eight respects in which new social movement theory was useful in our investigation. These are firstly, the European theoretical tradition and secondly, the mediating role of social movements between the abstract world of academia and politics. Thirdly, the role of the "old" labour movement and fourthly, the question of extra-parliamentary political activities. Fifthly, NSM's concern with political attitudes of people; sixthly, the role of environmental movements, as one type of NSM; seventhly the relationship between social movements and the media and finally, the composition of NSM participants. We will look at these in turn.

Firstly, on the theoretical level, an important characteristic of NSM theory is its European theoretical tradition, based on Marx's and Weber's writings. This is very similar to the eastern European social scientific tradition which is also based on the philosophical, political and historicist approach, as is the western European tradition, and which includes Marxism which was influential in the Soviet Union and both Marxism and a Weberian approach which were influential on Hungarian sociology.

Secondly, the mediating role between the abstract world of academia and the practical world of politics, which became a significant feature of NSMs, was similar to what occurred in movement activities in eastern Europe in the late 1980s and the 1990s. Social movement activists were often scientists who became very active politically and played an important role in social movements, including environmental movements, in which natural scientists played a particularly important role.

Thirdly, the role of the "old" labour movement was similar in the sense that it has been rejected in western Europe and also in eastern Europe for its old-fashioned approach has been replaced by demands for new politics. There were obvious differences as well, mainly that this "old" labour movement was in power in eastern Europe. However, eastern European peoples' dislike of the Communist/Socialist parties was hugely exaggerated in the western press, as proved by their recent election victories in many eastern European societies. The "renewed" old labour movement in the form of a modernized socialist party is popular all over eastern Europe. However, parallel to this, there is also a general disillusion with political parties and a growing interest in "new" political actions, which is similar to what is argued by NSM theorists.

This leads to our fourth point, the question of extra-parliamentary political activity. As suggested earlier, in the civil society section of this chapter and in previous chapters, in Hungary social movement activists are very conscious of their political role as lying outside the sphere of political parties, just as in western new social movements. Here, however, we have to distinguish between the two countries because this is not the case in Russia, where there is a lack of clarity in the sphere of the polity. This is due to the fact that political parties come and go with great frequency and do not fulfil their accustomed role in democracies. This leads to a lack of distinction between the opposition role within a party or a social movement, the emphasis being on opposition rather than on social movement versus political party.

Fifthly, new social movements in western European democracies embody three main sets of attitudes: (1) anti-consumptionism and postmaterialism, which originated in the 1950s and 1960s period of economic boom, (2) demands for autonomy and identity, and (3) a rejection of centralization, control and bureaucratization. Whereas anti-consumptionist and postmaterialist attitudes are mostly absent in eastern Europe, due to the lower level of economic development compared with western Europe, the pro-autonomy, anti-bureaucracy, anti-centralization and control attitudes are strongly present both in Hungarian and Russian societies and environmental movements. Thus emphases advocated by new social movement theorists, such as Offe (1985) and Habermas (1981; 1992) in particular, are of specific use for researchers in the eastern European region.

The sixth point refers to the fact that both in Russia and Hungary environmental movements became one of the most popular types of social movements and this is connected to western European new social movements. The recognition of the state of the environment as a major problem is a direct result of influence from the West in both cases. Although this direct inspiration arrived in Hungary a lot earlier than in Russia, where it only influenced public opinion a lot later, in the glasnost period, today western environmentalist ideas are equally influential in both countries.

The seventh respect in which new social movement theories have been proven relevant concerns the movements' relation to the media. The weight of public opinion and the role of mass media in influencing it is well recognized by both western NSMs and eastern European social movements. The media's ability to extend the movements' domain to a wide circle of people is well understood everywhere. This recognition is similar in both countries but there is a difference between the way social movements can use the media to help their own cause. Again, it is Hungary, where the media is independent enough to be used by the movements in a similar manner to the western cases, by securing media presence at demonstrations and contacting journalists to "advertise" movement activities. In Russia, however, the overall political control and lack of democracy prevents the media from acting similarly to the Hungarian or the western cases, as discussed earlier. However, the international media outside Russia do play a role to a limited degree, which facilitates some recognition of the movements even though this can only create a small and indirect influence on public opinion within Russia and thus achieves very little of its original intention.

Finally, concerning the composition of social movement participants, new social movement theorists' recognition that most participants in new social movements come from the most educated part of society was borne out in eastern Europe. Although the distinction between private and public sector employment was irrelevant at the time when environmental movements emerged in Russia and Hungary and seems to be irrelevant even now when the proportions are changing to some extent, the educational level of the activists is very important in both societies. Most core movement members and leaders come from educated groups and are often middle class by origin (if we can use this term in eastern Europe at all). They are mostly natural scientists by origin in the case of environmentalists rather than social scientists as in other types of social movement. There is, however, a wider mixture of the different "classes" in eastern Europe than in similar western movements. This is related to two important factors. Firstly, that labour movements, or "old" movements, against which new social movements originally emerged, play a somewhat different role in eastern Europe from the point of view of their class construction. These were highly institutionalized movements which did not have a majority of working class people as members, but were made up of party apparatchiks, whose fathers might have been working class, and were more likely to be careerists themselves than devoted revolutionaries. Secondly that, although state socialist countries did not become classless societies by any means, the kind of rigid class division which characterized western Europe at the time of the emergence of new social movements was certainly not similar to the eastern European situation. This, in fact, makes class comparison so difficult that it is safer to talk about the level of education when comparing the eastern and western situation, in which case Russia and Hungary are fairly similar to the western cases. It is, however, important to notice that in terms of the age of movement activists, the two countries differ from each other. In Hungary the age structure of environmental movement participants is similar to that in western Europe: it is dominated by the middle aged and the student-aged, while in Russia many older people of retirement age participate alongside the middle aged.

Summarising the relevance of new social movement theories in eastern Europe, we found that there are many elements of the new social movement theory which stand up when measured against an eastern European context. Thus the theory which deals with European matters unwittingly applies to the eastern part of Europe as well. There are significant differences, however, between Russia and Hungary, the latter being a society which has been exposed to western influence much earlier than Russia. Thus the new social movement theory's European and macro-sociological approach proved to be mcuh more relevant than any of the any earlier discussed, American born, theories in explaining our cases. As we have demonstrated, NSMs had an important impact especially in Hungary but later in Russia as well. The numerous similarities between the two parts of Europe thus made this fundamentally European concept very relevant in the eastern European context as well.

The NSM theory emphasizes the importance of political challenge in the social and cultural changes. This is where it substantially differs from the political opportunity structure perspective, which focuses on the political context instead.

The Relevance of the Political Opportunity Structure Theory

As we have argued earlier, the political opportunity theory developed following the arguments of the more macro-oriented group of theorists many of whom originally were labelled as RM theorists. The need to examine macro structures when analyzing social movements thus arose in

America and has been the cornerstone of the European investigations. The gap between the two approaches was recognized by Klandermans and Tarrow (1988) who expressed a strong wish to bridge it, suggesting that both RM and the European approach had a lot to offer and should be synthesized. This desired synthesis, however, turned out to be difficult to achieve due to the incompatibility of the different approaches which derive from different political cultures.

The political opportunity structure theory, although it does not synthesize two existing theories, managed to bridge the gap by providing a sufficiently general framework to be successfully used in any existing society. It abandoned the particularity of all the previous theories and is neither "American" nor "European", but "universal". Thus we should not be surprized that this theory was found useful in understanding our cases. Accepting the theory's fundamental argument that it is the political opportunity structure which is responsible for the emergence and effect of social movements in different societies, we can explain the diversity within eastern Europe as well. In this sense eastern Europe is truly part of the European tradition: the different countries provide varying political scenarios and those in turn lead to specific patterns of social movement development.

This is very much the case at present time but was, in fact, true even in the socialist period. Under the "uniform" political structure there was a range of cases because there were important differences in each individual society within the so-called Soviet bloc. There were two countries where the church became an important centre of the opposition: Poland (the Catholic church) and East Germany (the Lutheran church). A strong, nationwide social movement (Solidarity) developed in Poland, while no movements could emerge in Romania and Albania due to the most repressive party politics and extensive informant system. There was some clandestine movement formation in Bulgaria and a mainly intellectual movement (Charter 77) in Czechoslovakia. There was a high degree of political repression in the Soviet Union but opposition emerged in different shapes and forms from religious and nationalistic to workers' resistance or even pop groups. And, finally, there was a large degree of political tolerance in Hungary but social movements were not a frequent phenomenon in the socialist period. Thus in each society a different political opportunity structure was combined with a distinct national historical "inheritance" leading to different chances for social movement activities in various cases (for more on this, see Pickvance, 1998a).

In a comparative study, the political opportunity structure theory is doubly useful. It helps us to understand why social movements could develop more successfully in one society compared with another, and to explain why the two cases follow such different paths. As we argued earlier, in analyzing the available intellectual resources of Russian and Hungarian environmental movements, we found so many similarities that the fundamentally different outcomes of these movements remained unexplained and even run into contradictions. We have found, for example, that Russian movement activists are often almost fanatical concerning their "mission" and devote considerably more time and energy to the movement than their more practical Hungarian counterparts. Russian movements are also much more structured than Hungarian ones whether we look at individual movements, city or national federations. Following the RM arguments, a well structured movement organization is certainly the basis of a successful social movement. Yet success eludes Russian social movements whereas it is a typical outcome of movement activity in Hungary. Loosely structured Hungarian movements make more progress in every sense than well structured Russian ones. What is behind this fundamental contradiction? Why cannot resource mobilization theory offer an explanation? The answer is that the cause of the different patterns of social movement development lies outside the "movement-organization". It depends on the different political context of the two societies in which these movements have to operate. Thus it is the political opportunity structure theory which offers the explanation which can help us find the answer to this crucial question.

The political opportunity structure theory draws attention to several important aspects of social movement development in each society. These include the degree of openness/closure of formal political access, the degree of stability/instability of political alignments, the availability and strategic posture of potential alliance partners and the political conflict within and between the elites. All these aspects concern relations between social movements and the polity. Thus we turned our attention to the analysis of the social movements' relationship with the authorities, the link between movements and polity.

Having analyzed step by step the state of the polity and the authorities' relationship with the movements at different levels, we concluded that there is a very limited openness of formal political access in the Russian case in great contrast with the much more open and cooperative system in Hungary. We found that there is a high degree of stability of political alignments in Hungary and a largely unstable political party formation in Russia and that potential alliance partners are available for social movements in Hungary up to the highest level of politicians. This, again, is in contrast with the Russian case where there is very little availability of potential allies and when such cooperation exists it is with the weakest link in the fragile political system, the mainly powerless local councillors. However, we also found that there are frequent political conflicts within the elite and between elites in both countries.

Political opportunity structure theory also draws attention to the configuration of power and informal procedures and dominant strategies, including aspects of the strategies of the authorities. These were very different in Russia and Hungary. In the Russian case power is held by unelected officials in authorities. Neither formal procedures nor informal contacts are helpful because the authorities are obstructive towards social movements. Russian movements are not even in a position to be pro-active in building relationships with authorities. The situation in Hungary is just the opposite. There is a cooperative relationship between authorities and movements with mutual respect. In Hungary many social movement demands are facilitated by the authorities because the movements are respected as political actors with influence on local public opinion, hence their chances of success are high. In Russia social movements suffer a high level of repression and their demands are not facilitated by the authorities who do not even negotiate with them. Consequently they are isolated and their chances of future success are remote and diminishing. Hence the strategic options of "challengers" are very limited.

Thus the context for environmental movements is becoming very different in the two societies. While in the late 1980s and early 1990s the political situation, as described earlier, was similar in many ways it is diverging as time goes on. Hungarian movements have established themselves and enjoy the support and cooperation of the authorities. They survive and are achieving substantial results on all fronts. Russian movements, on the other hand, are becoming isolated and disillusioned, which persuades many of them to give up their activities.

The gap between the two societies is also growing. In Russia we found a closed political system which is confrontational towards challengers and is not ready to listen and compromise, or to allow access to policy-making or be influenced by outside opinion. There is no mediation between authorities and environmental movements. Even basic requirements are missing, such as an independent legislative system to allow social movements to seek justice. However, in contrast to Kitschelt's suggestion (1986), this closed and hostile political system did not push social movements into a confrontational mode; they did not become more aggressive. Instead it led to their weakening and steady disappearance from the political scene. This is because Kitschelt's model fails to distinguish between the structural and the contingent features of political systems, such as the strategies and tactics of other political actors, including the absence of effective opposition, as argued by Rootes (1992). The lack of belief in Russia's ability to create a democratic society is the saddest outcome of this situation.

Kitschelt was, however, right to argue that open political systems invite non-confrontational, assimilative strategies towards social movements, which is exactly what happened in Hungary. The many access points within the public sphere, including the authorities and social movements, created a healthy pluralistic political structure in which the mediation between social movements and executives is an ongoing process. There is an independent legislative system which provides a fair chance for social movements when needed. In one case it was the government itself which provided the finances for the movement to be able to sue a highly polluting former state company. The openness of the polity leads to compromises in many cases which means that the environmental movements not only achieve concrete policy changes but more importantly increase their reputation and respect among both the authorities and the public. This has created a political consensus in which most sides do their bit towards maintaining a democratic system in Hungary.

In sum, the political opportunity structure theory was found exceptionally useful in our analysis. The state of the environmental movements could not be explained by an analysis of the movements themselves. It was necessary to take into consideration the political context and the interaction between the polity and movements. Thus societal context was the most important explanatory variable when understanding the behavior of environmental movement activists. The analysis of resources and organizational aspects did not lead to satisfactory explanations. Thus the political opportunity structure theory proved to be as useful in interpreting what is happening in eastern Europe as it has elsewhere. Finally, we will turn to the most recent theory which we looked at when analyzing environmental movements in Russia and Hungary.

The Relevance of the Cognitive Approach

The main argument of the cognitive approach is that the most important role of social movements is not only to achieve concrete goals but to disseminate a certain "set of knowledge" within "social space", that is national or global society.

The concept of environmental consciousness was important in our investigation for three reasons. Firstly, because, as Eyerman and Jamison (1991) found in their studies of Sweden and the Netherlands, we found that Russian and Hungarian environmentalists perceived environmentalism as a subject which has long term consequences and spreads far beyond their locality. Subjects such as the resolutions of the Rio conference were often brought up and were taken very seriously. Thus the global nature of green matters was central to their thinking. Secondly, both Russian and Hungarian environmentalists considered it important to disseminate green views to a wider circle, beyond their own group. In Hungary, however, this wish has not only has been articulated but has also been converted into action and most movements actively disseminate environmental knowledge to a wider circle and especially among the younger generation. Russians, on the other hand, only mentioned the idea of dissemination of environmental views as a desirable aim. Russian environmental activists frequently mentioned that they saw the lack of environmental consciousness among their fellow citizens as a major problem and that environmental issues should be thought of globally while acting locally, but had done nothing to change the situation. None of the groups or even individuals had approached people, for example younger ones, to plant environmental concerns in the thinking of the new generation. Not even during the peak period of the late 1980s, when so many Russians became concerned with environmental problems, was there an attempt to spread environmental consciousness among the young. The wish, however, remained there even at the time when green movements were in a decline.

Thirdly, the cognitive consciousness theory argued that a movement's success cannot only be measured by its concrete achievements. Rather the way they manage to carve out new knowledge and a new understanding of issues, in our example environmental ones, within the society they act in is also important. We have to emphazise at this stage that the movements themselves also articulated such views. The movements measured their own success by the extent to which they could change public opinion, locally or nationally, concerning the issues of particular concern to them and about environmental matters generally. We also incorporated this aspect in our judgement of success or failure and concluded that Russian environmental movements were very unsuccessful and Hungarians were successful in this respect. The examples of the Swedish and Dutch models were also useful from a comparative perspective. The centralized, top-down Swedish model, in which political openness is limited selectively and radical environmentalism is not successful but the traditional nature-loving conservationalism combined with a strong desire and belief in technical development, has a lot in common with the Russian case. Even though the Swedish scenario is based on an economically sound and well organized society, both of which are lacking in Russia at present, future economic development could lead eventually to a Swedish-type of development, as far as environmentalism is concerned.

The Hungarian case is similar to the Dutch "model", as described by Eyerman and Jamison (1991), in the sense that both political parties and grassroots groups play an important role in the political context. Green parties, however, are politically weak and extra-parliamentary, professionalized environmentalist groups have become the leading actors in environmental issues and both public education and lobbying are present among the tactics used by the movements.

Conclusion

In this chapter I have evaluated a number of existing theories, including the theory of civil society, collective behavior, resource mobilization, new social movement, the opportunity structure theory and the cognitive approach. My aim was to establish whether these theories, all of which were developed in a western context, had any relevance in our eastern European cases. As I have demonstrated in this chapter the different theories were of varying relevance in our cases. The two American theories, the collective behavior and resource mobilization approaches, both of which had been developed decades earlier, were found least useful in our investigations. But the main problem with these two theories was not their country or time of origin. Rather it was that their arguments were found less useful than those of other theories. Nevertheless, even these two theories were relevant to a certain degree even though they could not help us in answering the most important questions we posed: why were Russian movements were developing so differently from Hungarian ones and why were they were they so much less successful than environmental movements in Hungary.

The other four theories were all very relevant in our investigations. But again, some more so than others. New social movement theory brought out many similarities, partly stemming from the fact that it is "European". The cognitive approach was useful both in its arguments and in comparative terms. The two most useful theories, however, turned out to be civil society and political opportunity structure theory. This is because I came to the conclusion that it is the political context, and especially the degree of development of democratic institutions, which is the single most important key to understanding the differing evolution and success of environmental movements in the two countries. The differences in political context in Hungary and Russia also have an effect on the level of resource availability and the degree of media support which we showed to have a significant influence on environmental movements. Thus the relevance of the civil society and political opportunity structure theories is greatest because they focus on the crucial and fundamental aspect of the analysis of social movements, showing their embeddedness within society.

Finally, the fact that existing theories are relevant implies that Hungary and Russia are part of the same "analytical universe" as those societies in the "west" which inspired them. They differ from each other but so do "western" societies. "Universal" concepts, such as those dealing with problems of democracy, citizenship, social movements can be applied to analyze social and political processes in any society but do not require that these processes be identical. In the next chapter we turn to our final conclusion.

Note

1. *Medvetánc* was a periodical published by the KISz organization in one of Budapest's main Universities, the Eötvös Lórand University.

11

Conclusion

The focus of this book was the development of environmental movements in two former Soviet bloc societies. In this concluding chapter I shall attempt to bring together some of the disparate threads of this study. I outline the main aims of the book and then summarise its contribution to the theory of social movements. Thirdly, I will outline an understanding of political changes in eastern Europe including a number of questions for future research and finally discuss the likely future development of environmental movements and democratization in eastern Europe.

Main Aims of the Book

In the introduction I set out four major aims. These were: to carry out an in-depth study of environmental movements and authorities in two eastern European societies using an over time research design; to carry out a systematic comparison of the two societies; to examine the relevance of existing theories of opposition in Soviet-type societies, civil society theories and social movement theories and to study the relevance of historical continuity in these two societies.

At the end of my study how far have these aims been achieved? In relation to the first aim I studied environmental movements and their interaction with authorities in Russia and Hungary after the regime change. This is an advance on the many previous studies of social movements which have not carried out empirical research on authorities in relation to social movements. In my view it is crucial to study empirically both parties to the relationship between social movements and authorities in order to understand the interaction.

Secondly, I carried out a systematic comparison of the two societies and their environmental movements. Previous writers on eastern Europe have mostly done empirical studies in one country and have not made systematic comparisons. In my opinion a comparative analysis is essential to bring out empirical similarities and differences, and to help in developing theories about them.

Thirdly, we have given considerable attention to examining the relevance of existing theories. I have considered theories relating to opposition in soviet-type societies, civil society theories and social movement theories. The theories of opposition helped me understand the very different experiences of regime opponents in these two societies and also the subsequent development of democratic opposition once this was possible. Civil society theories were useful in demarcating the sphere of political activities outside the state and political parties, and social movement theories have addressed the pattern of emergence and success of movements which were the focus of our study.

Contribution to Theory

The book contributes to theory in three fields: theories of opposition in Soviet-type societies, theory of civil society and social movement theories. Previous writers on opposition in Soviet-type regimes have classified it in various ways but usually in terms of a single dimension. My own contribution is to introduce two dimensions of opposition. The first of these refers to the level of resentment against the regime and the distance from the ideas of the ruling Communist Party. The second dimension referred to abstract and concrete types of demand.

In relation to theories of civil society, I agree with those writers who regard civil society as an over-used and often poorly defined term. Once clearly defined, however, it is a useful concept to separate grassroots activities from political parties and other parts of the formal political structure, though the connection between them is obvious and important. But the aim of the concept of civil society is not to explain why and how social movements develop. This is the task of social movement theories.

In the previous chapter I showed that, despite their western origin, the concepts in a number of existing theories of social movements were useful in understanding eastern European experience, though to varying degrees. This shows that they have explanatory power in a greater range of situations than their authors envisaged. It also shows that eastern Europe is not an isolated entity which requires a totally unique conceptual approach. At the broadest level, the political opportunity structure theory proved most useful. This justifies our decision to study authorities empirically. However, the other theories were all valuable for understanding specific aspects of social movements. Although my conclusions are based on the study of one type of movement, in the light of other research (including my own on housing movements), I believe that they are not limited to environmental movements.

Contribution to an Understanding of Eastern Europe

A major conclusion of my analysis has been to emphasise the contrast between Hungary and Russia. I first showed this in my discussion of the historic development of the two societies. This discussion emphasized the distinctiveness of the pre-socialist and socialist periods, and showed that these periods had a major impact on the process of democratization and the emergence of political opposition. I see the very different levels of democratization as the key to the contrasting social movement experience in the two societies.

I also showed that Hungarian environmental movements were strongly influenced by the German and Austrian models whereas in Russia western influence is combined with the more romantic tradition of national nature protection. The recent evolution of environmental movements in Hungary and Russia since the regime change has also been very different. At the "peak" of the transition there was an upsurge of social movements in both countries and a sudden growth of interest in environmental issues. This resulted in a sharp increase in environmental movements. While this was very similar in both countries the subsequent decline was different.

In Hungary, as in other non-eastern European societies in transition, such as Spain and Portugal, a certain decline of movement activities was due to political stabilization with the establishment of well functioning political parties and the successful restructuring of authorities. However, a fairly steady level of environmental movement activity continues to exist. In Hungary the more favourable environment and achievement of considerable success has led to a stabilization of environmental movements at a higher level. In Russia, however, neither of these conditions was present: no system of functioning political parties developed and the frequent reorganization of authorities did not result in a more democratic system. This different societal context is responsible for the difference in environmental movements. Failure to achieve their aims and the repressive political environment led to a sharp decline in environmental movements in Russia.

Finally, I need to ask whether my conclusions about the development of democratization in Hungary and Russia are influenced by the choice of environmental movements as the focus of research. In both societies environmental activism did develop in the final years of state socialism and undoubtedly to some extent even contributed to the development of political pluralism. This was because at the end of the socialist period environmental movements played a double role in both societies. They were tolerated by the outgoing regime and therefore became the focus of opposition. Many politically active people joined environmental movements in order to express their cautious discontent with the regime.

Later, when opposition was legalised, these primarily politically oriented people left environmental movements and joined political parties. Environmental movements were thus the first organized political activity in which these people gained experience as political activists. However, with the change of regime, environmental activity lost its privileged character as a much wider range of types of political participation became possible and environmental movements became only one type of social movement. The implication of this is that since the change of regime (but not before), environmental movements can be considered typical of social movements of all kinds.

Despite the achievements of this study I do not consider that it has at all exhausted its field. There is clearly scope for similar studies in former state socialist societies other than Hungary and Russia, and also for a comparison between former state socialist societies and Mediterranean and Latin American societies that have undergone transitions from authoritarian rule. The role of enterprises and their influence (or the lack of it) on the emergence of environmental awareness also deserves explicit attention. There is also a need for systematic national studies (including developed western societies) and data archives which would be a great value to those undertaking in-depth studies of a limited number of movements.

The Future Development of Environmental Politics and Democratization in Eastern Europe

In order to discuss the future of environmental movements one needs to consider the forces that favour their development. Firstly, the "objective" level of environmental ills is relevant. Though industrialization in the past and the more recent lack of attention to environmental problems has led to serious environmental damage, the first question is whether this damage is recognized and secondly to what extent grievances are translated into action.

As a result of the upsurge of interest in environmental problems at the time of the regime change, one would expect a widespread recognition of their existence in both societies today. In Hungary this recognition does exist among ordinary people and there is continuing interest in environmental issues. In Russia, however, where the "objective" situation is worse than in Hungary due to the fact that industrialization was more developed, the earlier recognition has given way to the "official" ideology that the first priority is to get the economy working properly, and that the expenditure of energy and resources on green matters is a "luxury". However, at the central governmental level there is strong evidence that environmental policy is given a low priority in both countries (with the exception of the mayor of Budapest who pursues the matter).

It could be suggested that in Russia there will be a greater increase of activism based on the objective situation and the failure of the political leadership to respond to demands. But, as the theoretical literature on social movements suggests, grievances are much less important than the political context and the level of consciousness in determining activism. It is these social forces which are more likely to shape the future pattern of environmental movements. In Russia there is a "closed" polity; central and local government are not open to outside suggestions, pressure or any kind of opposition. This is likely to lead to the continuing decline of grassroots political activities including environmental movements. The lack of success of current environmental movements and the hostile attitude of the authorities will continue to discourage people from spending time and energy on "futile" activities such as social movements, including environmental movements. I also think that environmental policy will continue to be given a very low priority by the government.

The divergence between the two societies is likely to continue. In Hungary, where central and local government are much more open to outside influence, civil society will survive and social movements will continue to have a say in politics. Environmental movements will maintain a steady level of activity. Many "old" movements will continue to exist and new ones will come into existence all over the country. The relative success of present day environmental movements will provide a good example for the new ones as to how to cooperate and compromise in order to survive. Thus environmental issues will remain fairly important in people's minds and in the media. Whether environmental policy will be given a higher priority by the government remains to be seen.

Appendix

The research for the book was conducted between 1989 and 1995. After some preparation it was at the time of the March 1990 general elections that I first visited Hungary in order to observe the first free elections, including the campaign. I visited political party headquarters, collected election leaflets and manifestos, talked to people, and followed the election campaign methods on television. At the same time I also started my interviewing of representatives of environmental movements. This was before the start of the research project funded by the Economic and Social Research Council (ESRC), the British Government funding body. I visited and interviewed present and former activists and leaders of several environmental movements including the Danube Circle, Green Future, the Free Democrats Green Group, the Eotvos Lorand University's green group (which later became one of the founders of the Air Group). Apart from interviewing activists individually, I also participated in several meetings and conducted participant observation in order to establish the basis for subsequent over-time interviewing.

The ESRC started to fund our research in 1991. We prepared two separate structured interview schedules: one for social movement participants and another for authority members. These were carefully piloted by the local collaborators in both countries. Having analyzed the pilot interviews (both in Russian and Hungarian) we invited our collaborators to Britain and I had a training session with the interviewers in order to get the maximum results and consistency. The interviews using the provided schedules were conducted by collaborators in Russia and Hungary. The collaborators sent over the interviews on a continuous basis and I monitored them very carefully in order to maintain consistency.

I also went to Hungary and Russia several times to conduct further interviews with local and national authority members, movement leaders and activists as well as to participate in meetings, monitor the media and learn about the political development in both societies on a regular basis.

I conducted twenty-three intensive interviews with environmental movement activists (both leaders and rank and file members) and local and national authorities which included elected representatives, and local and national government officials, and our local collaborators conducted a further 106 interviews, closely following the interview schedules with which I provided them. (The ESRC project also included another 187 interviews conducted by the local collaborators which will be the subject of another book: Láng-Pickvance, K., Manning, N. and Pickvance, C., *Citizen Action in Eastern Europe*, Oxford University Press, forthcoming in 1999).

The movements selected here were representative of the types of movements which existed in each country. Thus in Hungary, where there are no federal organizations, the movements which I call "national" are of national importance, because of their subjects. These always have their headquarters in Budapest, representing the dominance of the capital in Hungary. There are, however, important local movements inside and outside the capital which were also represented in my choice.

In Russia, on the other hand, the importance of federal organizations made me decide to include them. Russian local movements do not usually become as important as national ones, nevertheless it was important to include them in order to represent the variety present in Russia. Here too Moscow plays a central role in political life but of course federal organisations stretch beyond Moscow.

In order to conduct a systematic comparison of Russia and Hungary I used the same detailed interview schedules in both countries. This, combined with my own interviews which did not use the schedule, worked very well. The topics covered in the interviews with movement participants were: the history of the movement, grievances, change in the political situation, institutional support, the goals of the movement, participants, organizational aspects, structure, size, motivation of participation, leadership, funding and other resources, contacts with political parties and other movements, conflicts, relation with the media, attitude of population to the movement, success, dynamics.

Topics covered in the interviews with the authorities covered: clarification of the authorities' functions, role of representatives and officials, relations with other authorities and movements, how they obtained information and formed their views, what official relationships and personal feelings the authority members developed towards the movements.

As mentioned earlier, in 1989, when political changes occurred in eastern Europe I decided to turn my academic interest back to Hungary and the Soviet Union and visited the region in order to explore the new situation. My interest in environmental issues was also very strong even prior to this time which is why I decided to focus on green issues.

Later, when the ESRC launched the so-called "East-West" initiative in order to promote research in the area, three of us (Nick Manning, Chris Pickvance and myself) decided to apply for a grant which we were awarded in 1991.

The aim of the ESRC project, as opposed to my agenda, was to conduct research in two republics of the Soviet Union (Estonia and Russia) and in Hungary in order to investigate housing and environmental movements in three locations during a three year period (1991-1994). In addition to the interviews with environmental movement activists, referred to above, within the project we conducted in-depth interviews with housing movement activists and, in relation with housing movements, with authority members. Within the ESRC project we also conducted three large scale surveys in three locations: Estonia, Hungary and Russia (a sample of 2,000 people in total) in order to gauge ordinary people's attitudes concerning their housing problems, environmental and general political issues and to establish to what extent they were willing to participate in political activities. These will be the basis of our book, mentioned above. In this book I focus on the question of democracy. Environmental movements provided a good way of analyzing how and to what extent democratic intitutions developed in these two post-Soviet societies.

Bibliography

Acton, E. 1992. *Rethinking the Russian Revolution*. London: Edward Arnold.

Almond, G. A. and Verba, S. (eds) 1989. *The Civic Culture Revisited*. London: Sage Publications.

Anderson, M. S. 1969. *Peter the Great*. London: Thames and Hudson.

Andreyev, N. 1976. Appanage and Muscovite Russia, in R. Auty, D. Obolensky and A. Kingsford, *An Introduction to Russian History*. Cambridge: Cambridge University Press.

Arató, A. 1991. Social movements and Civil Society in the Soviet Union, in J. B. Sedaitis and J. Butterfield (eds) *Perestroika from Below: Social Movements in the Soviet Union*. Boulder: Westview.

Arató, A. and Cohen, J. L. 1992. *Civil Society and Political Theory*. Cambridge, Mass.: MIT Press.

Arendt, H. 1973. *The Origins of Totalitarianism*. Brighton: Harvest Publishers.

Armstrong, J.A. 1961. *The Politics of Totalitarianism*. New York: Random House.

Bak, J. 1990. The late mediaeval period, 1382-1526, in F. P. Sugár (ed.) *A History of Hungary*. London: I. B. Tauris.

Bárány, G. 1990. The age of royal absolutism, 1790-1848, in F. P. Sugár (ed.) *A History of Hungary*. London: I. B. Tauris.

Barr, N. and Harbison, R. W. 1994. Overview: hopes, tears and transformation, in N. Barr (ed.) *Labor Markets and Social Policy in Central and Eastern Europe*. Oxford: Oxford University Press.

Barta, I. 1975. Toward bourgeois transformation, revolution and war of independence (1790-1849), in E. Pamlényi (ed.) *A History of Hungary*. London: Collets.

Bater, J. H. 1987. St. Petersburg and Moscow on the eve of the revolution, in D. H. Kaiser (ed.) *The Workers' Revolution in Russia, 1917: the view from below*. Cambridge: Cambridge University Press.

Batkay, W. M. 1982. *Authoritarian Politics in a Transitional State, 1919-1926*. New York: Columbia University Press.

Besancon, A. 1978. *The Soviet Syndrome*. New York: Harcourt Brace Jovanovich.

Berend, T.I. 1990. Contemporary Hungary, 1956-1984, in F. P. Sugár (ed.) *A History of Hungary*. London: I. B. Tauris.

Berezhovski, V. N. 1990. *Neformalnaja Rossija*. (Informal Russia) Moscow: Molodaja Gvardija.

Bibó, I. 1979. *Selected Essays*. (edited by P. Kende) Paris: Dialogues Européens.

Bilocerkowycz, J. 1988. *Soviet Ukranian Dissent: a Study of Political Alienation*. Boulder: Westview.

Bobbio, N. 1988. Gramsci and the concept of civil society, in J. Keane (ed.) *Civil Society and the State*. London: Verso.

Bogoslovskij, M.M. 1963. Peter's program of political reform, in M. Raeff (ed.) *Peter the Great: Reformer or Revolutionary?* Boston: Heath.

Bottomore, T. 1979. *Political Sociology*. London: Hutchinson.

Boyce, J.H. 1993. Local government reform and the new Moscow City Council. *The Journal of Communist Studies* 9(3): 245-271.

Bozóki, A. 1988. Critical movements and ideologies in Hungary. *Südosteuropa*, 7-8.

Brzezinski, Z. (ed.) 1969. *Dilemmas of Change in Soviet Politics*. New York: Columbia University Press.

Bugajski, J. and Pollack M. 1989. *East European Fault Lines: Dissent, Opposition and Social Activism*. Boulder: Westview Press.

Burnham, 1941. *The Managerial Revolution*. London: Putnam.

Butterfield, J. and Sedaitis J. B. 1991. The emergence of social movements in the Soviet Union, in J. Butterfield and J. B. Sedaitis (eds) *Perestroika from Below: Social Movements in the Soviet Union*. Boulder: Westview.

Butterfield, J. and Weigle M. 1991. Unofficial social groups and regime response in the Soviet Union, in J. Butterfield and J. B. Sedaitis (eds) *Perestroika from Below: Social Movements in the Soviet Union*. Boulder: Westview.

Brand, K.W. 1990. Cyclical aspects of New Social Movements: waves of cultural criticism and mobilization cycles of new middle-class radicalism, in R. J. Dalton and M. Kuechler (eds) *Challenging the Political Order*. Cambridge: Polity Press.

Breslauer, G. W. 1980. Khrushchev reconsidered, in S. F. Cohen, A. Rabinowitch and R. Sharlet (eds) *The Soviet Union since Stalin*. Bloomington: Indiana University Press.

Brown, A. 1984. Political power and the Soviet state, in N. Harding (ed.) *The State in Socialist Society*. London: Macmillan/St. Anthony's College Oxford.

Bryant, C. G. A. 1993. Social self-organization, civility and sociology: a comment on Kumar's "Civil Society", *British Journal of Sociology* 44(3): 397-401.

Brzezinski, Z. 1966. *Totalitarian Dictatorship and Autocracy*. New York: Praeger.

Brzezinski, Z. 1969. The Soviet political system: transformation or degeneration? in Z. Brzezinski (ed.) *Dilemmas of Change in Soviet Politics*. New York: Columbia University Press.

Campeanu, P. 1988. *The Genesis of the Stalinist Social Order.* London: M. E. Sharpe.

Carr, E. H. 1979. *The Russian Revolution: From Lenin to Stalin, 1917-1929*. London: Macmillan.

Carson, R. 1965. *Silent Spring*. Harmondsworth: Penguin.

Cohen, G. B. 1989. The social structure of Prague, Vienna and Budapest in the late nineteenth century, in G. Ránki (ed.) *Hungary and the European Civilization*. Budapest: Akadémiai Kiadó.

Cohen, J. L. 1986. *Class and Civil Society: The Limits of Marxian Critical Theory*. Oxford: Oxford University Press.

Cohen, S. F. 1985. *Rethinking the Soviet Experience*. Oxford: Oxford University Press.

Conquest, R. 1969. Immobilism and decay, in Z. Brzezinski (ed.) *Dilemmas of Change in Soviet Politics*. New York: Columbia University Press.

Cracraft, J. 1988. Opposition to Peter the Great, in M. Mendelsohn and S. Shatz (eds) *Imperial Russia*. Northern Illinois University Press.

Crankshaw, E. 1978. *The Shadow of the Winter Palace: The drift to revolution 1825-1917*. London: Penguin Books.

Dahl, R. 1966. *Political Oppositions in Western Democracies*. New Haven: Yale University Press.

Dahl, R. 1971. *Polyarchy: Participation and Opposition*. New Haven: Yale University Press.

Dalton, R. J., Kuechler, M. and Burklin, W. 1990. The Challenge of new movements, in R. J. Dalton and M. Kuechler (eds) *Challenging the Political Order*. Cambridge: Polity Press.

Daniels, R.V. 1962. *The Nature of Communism*. New York: Random House.

Dawisha, K. 1988. *Eastern Europe, Gorbachev and Reform: The Great Challenge*. Cambridge: Cambridge University Press.

Deák, I. 1990. The Revolution and the War of Independence, 1848-1849, in F. P. Sugár (ed.) *A History of Hungary*. London: I. B. Tauris.

Deme, L. 1976. The radical Left in the Hungarian revolution of 1848. *East European Quarterly* 6(3): 126-141.

Deme, L. 1984. Hungary and the Habsburg monarchy, in G. Ránki (ed.) *Hungarian History—World History*. Budapest: Akadémiai Kiadó.

D'Encausse, H. C. 1981. *Stalin*. London: Longman.

de Tocqueville, A. C. H. C. 1946. *Democracy in America*. (De la Démocratie en Amerique, 1835) (Translated by H. Reeve) Oxford: Oxford University Press.

de Tocqueville, A. C. H. C. 1966. *The Ancient Regime and the French Revolution*. (L'Ancien Régime et la Révolution, 1856) London: Collins-Fontana.

Diani, M. 1992. The concept of social movement. *Sociological Review* 40: 1-25.

Diani, M. and Eyerman, R. (eds) 1992. *Studying Collective Action*. London: Sage Publications.

Djilas, M. 1957. *The New Class*. London: Thames and Hudson.

Doktorov, B. Z., Firsov, B. M. and Safronov, V.V. 1993. Ecological consciousness in the USSR: entering the 1990s, in A. Vári and P. Tamás (eds) *Environment and Democratic Transition: Policy and Politics in Central and Eastern Europe*. Dordrecht: Kluwer Academic Publishers.

Dukes, P. 1990. *The Making of the Russian Absolutism*. London: Longman.

Dunlop, J. B. 1983. *The Faces of Contemporary Russian Nationalism*. Princeton, N.J.: Princeton University Press.

Einhorn, B. 1993. *Cinderella Goes to Market: Citizenship, Gender and Women's Movements in East Central Europe*. London: Verso.

Elias, N. 1988. Violence and Civilization, in J. Keane (ed.) *Civil Society and the State*. London: Verso.

Engel, P. 1990. The age of the Angevines, 1301-1382, in F. P. Sugár (ed.) *A History of Hungary*. London: I. B. Tauris.

Erdei, F. 1978. A magyar társadalom a két háború között (Hungarian society between the wars), in *Szöveggyüjtemény a Szociológia szakositó hallgatói részére*. (Reader for sociology students) Budapest: MSzMP Budapesti Igazgatósága Szociólogia Tanszék.

Eyerman, R. and Jamison, A. 1991. *Social Movements: A Cognitive Approach.* Cambridge: Polity Press.

Ferro, M. 1985. *The Bolshevik Revolution.* London: Routledge & Kegan Paul.

Field, D. 1976. *The End of Serfdom.* Cambridge, Mass.: Harvard University Press.

Fischer, G. 1970. The intelligentsia and Russia, in C. B. Black (ed.) *The Transformation of Russian Society.* Cambridge, Mass.: Harvard University Press.

Fitzpatrick, S. 1979. *Education and Social Mobility in the Soviet Union.* Cambridge: Cambridge University Press.

Fitzpatrick, S. 1982. *The Russian Revolution, 1917-1932.* Oxford: Oxford University Press.

Fleischer, T. 1993. Jaws on the Danube: water management, regime change and the movement against the Middle Danube hydroelectric dam. *International Journal of Urban and Regional Research* 17: 429-443.

Frank, T. 1990. Hungary and the dual monarchy, 1867-1890, in F. P. Sugár (ed.) *A History of Hungary.* London: I. B. Tauris.

Frankel, J. 1992. 1917, The problem of alternatives, in E. R. Frankel, J. Frankel and B. Knei-Paz (eds) *Revolution in Russia: Reassessments of 1917.* Cambridge: Cambridge University Press.

Friedgut, T. 1979. *Political Participation in the USSR.* Princeton, N.J.: Princeton University Press.

Friedrich, C. J. and Brzezinski, Z. (eds) 1966. *Totalitarian Dictatorship and Autocracy.* London: Praeger.

Gábor, L. 1997. A környezetvédelem szervezete és eszközei Magyarországon. (Environmental legislation and the methods of environmental policy implementation in Hungary) Manuscript.

Galasi, P. and Sziráczki G. 1985. State regulation, enterprise behaviour and the labour market in Hungary, 1968-83. *Cambridge Journal of Economics*, 9: 203-219.

Gáti, C. 1990. From Liberation to Revolution, 1945-1956, in F. P. Sugár (ed.) *A History of Hungary.* London: I. B. Tauris.

Garrard, J. G. (ed.) 1973. *The Eighteenth Century in Russia.* Oxford: Clarendon Press.

Giddens, A. 1985. *The Constitution of Society.* London: Macmillan.

Goldstone, J.A. 1985. The weakness of organization: a new look at Gamson's *The Strategy of Social Protest, American Journal of Sociology* 5: 1017-1042.

Gorbachev, M. 1997. *Memoirs.* London: Bantam Books.

Grzybowski, M. 1991. The transition from one-party hegemony to competitive pluralism: the case of hungary, in S. Berglund and J. A. Dellenbrant (eds) *The New Democracies in Eastern Europe: Party Systems and Political Cleavages.* London: Edward Elgar.

Guardian, 1991. 27 August.

Gramsci, A. 1971. *Prison Notebooks.* (Edited and translated by Q. Hoare and G. Nowell-Smith) London: Lawrence and Wishart.

Habermas, J. 1981. New social movements. *Telos* 49: 33-37.

Habermas, J. 1992. *The Structural Transformation of the Public Sphere: An Inquiry into a Category of Bourgeois Society.* Cambridge: Polity Press.

Haimson, L. H. 1970. The parties and the state, in C. E. Black (ed.) *The Dynamics of Modernization: A Study in Comparative History.* London: Harper and Row.

Hammer, D. P. 1974. *The Politics of Oligarchy.* London: Praeger.

Hanak, P. 1984. Hungary's contribution to the monarchy, in G. Ránki (ed.) *Hungarian History—World History.* Budapest: Akadémiai Kiadó.

Hanson, P. 1993. Local power and market reform in Russia. *Communist Economies and Economic Transformation* 5(1): 45-60.

Haraszti, M. 1977. *The Worker in a Workers' State.* Harmondsworth: Penguin.

Haraszti, M. 1990. The Beginnings of Civil Society: the independent peace movement and the Danube movement, in V. Tismaneanu (ed.) *In Search of Civil Society: Independent Peace Movements in the Soviet Bloc.* London: Routledge.

Harding, N. 1984. *The State in Socialist Society.* London: Macmillan/St Antony's College, Oxford.

Hare, P. G. 1977. Economic reform in Hungary: problems and prospects. *Cambridge Journal of Economics.* 1: 317-333.

Haskó, K. 1988. A Demokratikus Politikai Rendszerek Működése és Fontosabb Intézményei (The functioning of democratic systems and their main institutions), in K. Haskó and K. Pais (eds) *Tanulmányak a Demokráciáról* (Studies in democracy). Budapest: Munkásakadémia Alapitvány.

Hegedüs, J. and Tosics I. 1996. Conclusion: past tendencies and recent problems of the east European housing model, in B. Turner, J. Hegedüs, and I. Tosics (eds) *The Reform of Housing in Eastern Europe and the Soviet Union.* London: Routledge.

Hegel, F. 1821 (1942). *Philosophy of Right.* Oxford: Clarendon Press.

Held, J. 1971. The heritage of the past: Hungary before World War I, in I. Volgyes (ed.) *Hungary in Revolution: Nine Essays.* Lincoln: Nebraska University Press.

Héthy, L. and Makó C. 1978. *Munkások, Érdekek, Érdekegyeztetés* (Workers, interests, conciliation of interests) Budapest: Gondolat Kiadó.

Heti Világgazdaság 1994a. Új Orosz Kormány: Hátra arc! (New Russian Government: About Turn!) 16(4)(766), 29 January, p.21.

Heti Világgazdaság 1994b: Kiválasztottak (Selected ones) 16(4)(766), 29 January, p.21.

Heti Világgazdaság 1995. Heten Vannak (They are seven), 17(3)(817), 21 January, p.24.

Hill, R. J. 1989. *The Soviet Union: Politics, Economics and Society from Lenin to Gorbachev.* London: Pinter Publishers.

Höensch, J. K. 1988. *A History of Modern Hungary, 1867-1986.* (Translated by K. Trayner) Harlow: Longman.

Hosking, G. 1990. *The Awakening of the Soviet Union.* London: Heinemann.

Hough, J. F. 1977. *The Soviet Union and Social Science Theory.* Cambridge, Mass.: Harvard University Press.

Hough, J. F. and Fainsod, M. 1979. *How the Soviet Union is Governed.* (Revised and enlarged edition.) Cambridge, Mass.: Harvard University Press.

Huntington, S. P. and Brzezinski, Z. 1964. *Political Power: USA/USSR.* New York: Viking Press.

Igrunov, V. V. 1989. Public movement: from protest to political self-consciousness, paper to ISA Conference, Moscow, October.

Ionescu, G. 1967. *The Politics of the European Communist States.* London: Weidenfeld and Nicolson.

Jamison, A., Eyerman, R. and Cramer, J. (with Laessoe, J.) 1990. *The Making of the New*

Environmental Consciousness: A Comparative Study of the Environmental Movements in Sweden, Denmark and the Netherlands. Edinburgh: Edinburgh University Press.

János, A. C. 1982. The Politics of Backwardness in Hungary 1825-1945. Princeton, N.J.: Princeton University Press.

Jenkins, J.C., 1985. The Politics of Insurgency: The Farm Worker Movement in the 1960s. New York: Columbia University Press.

Jessop, B. 1990. State Theory: Putting the Capitalist State in its Place. Cambridge: Polity.

Jeszenszky, G. 1990. Hungary through World War I and the end of the dual monarchy, in F. P. Sugár (ed.) A History of Hungary. London: I. B. Tauris.

Jones, R. E. 1973a. The Emancipation of the Russian Nobility. Princeton, N.J.: Princeton University Press.

Juhász, J., Vári, A. and Tölgyesi, J. 1993. Environmental conflict and political change: public perception on low-level radioactive waste management in Hungary, in A. Vári A. and P. Tamás (eds) Environment and Democratic Transition: Policy and Politics in Central and Eastern Europe. Dordrecht: Kluwer Publishers.

Kaiser, D. H. (ed.) 1987. The Workers' Revolution in Russia, 1917: The View from Below. Cambridge: Cambridge University Press.

Karády, V. 1989. Assimilation and schooling: national and denominational minorities in the universities of Budapest around 1900, in G. Ránki (ed.) Hungarian History–World History. Budapest: Akadémiai Kiadó.

Kassof, A. 1968. Prospects for Soviet Society. London: Pall Mall Press.

Katkov, G. 1967. Russia, 1917: The February Revolution. London: Collins.

Keane, J. (ed.) 1988. Civil Society and the State. London: Verso.

Keep, J. 1976. Imperial Russia, in R. Auty (ed.) Russian History. Cambridge: Cambridge University Press.

Kelly, H. 1986. The Social Structure of the Soviet Union. Cambridge: Cambridge University Press.

Killian, L. M. and Turner, R.H. 1972. Collective Behavior. Englewood Cliffs, N.J.: Prentice-Hall.

Király, B. K. (ed.) 1986. War and Society in East Central Europe: Vol.4.: East Central European Society and War in the Era of Revolutions, 1775-1856. Brooklyn: Brooklyn College Press.

Kis, J. 1988. A Filozófiai Intézettől a Beszélő szerkesztőségéig. (From the Institute of Philosophy to the Editorial Board of the Beszélő) (int. with E. Csizmadia) Valóság 12: 86-108.

Kitch, J. M. 1970. Eastern Europe 1918-44, in G. Schöpflin (ed.) The Soviet Union and Eastern Europe: a Handbook. London: Blond.

Kitschelt, H. 1986. Political opportunity structures and political protest: anti-nuclear movements in four democracies. British Journal of Political Science, 16: 57-85.

Klyuchevsky, V. O. 1968. A Course on Russian History. Chicago: Chicago University Press.

Kochan, L. 1983. The Making of Modern Russia. London: Macmillan.

Koenker, D. P. 1987. Moscow in 1917: the view from below, in D. H. Kaiser (ed.) The Workers' Revolution in Russia, 1917: The View from Below. Cambridge: Cambridge University Press.

Kolakowsky, 1977. Marxist roots of Stalinism, in R. C. Tucker (ed.) *Stalinism.* New York: Norton.

Komarov, B. 1980. *The Destruction of Nature in the Soviet Union.* New York: M.E. Sharpe.

Konrád, G. 1989. *Antipolitika: Az Autonómia Kisértése.* Budapest: Codex Rt. (English publication: 1984. *Antipolitics: an Essay.* (Translated by R. E. Allen) London: Quartet).

Kriesi, H. 1991. *The Political Opportunity Structure of New Social Movements: Its Impact on Their Mobilization.* Berlin: Wissenschaftszentrum fur Socialforschung.

Kriesi, H., Ruud, K., Duyvendak, J.W. and Giugni, M.G. 1992. New social movements and political opportunities in western Europe. *European Journal of Political Research,* 22: 219-244.

Kulcsár, L. and Dobossy I. 1988. Az ökológiai tudat és viselkedés társadalmi tényezŏi. Zárótanulmány. (The social factors of ecological knowledge and behaviour. Report) Budapest: Tömegkommunikációs Kutatóközpont.

Kumar, K. (ed.) 1971. *Revolution.* London: Weidenfeld and Nicolson.

Kumar, K. 1991. The Revolutions of 1989: Socialism, Capitalism and Democracy. Paper.

Kumar, K. 1992. The 1989 revolutions and the idea of Europe. *Political Studies* XL: 439-461.

Kumar, K. 1993. Civil Society: an inquiry into the usefulness of an historical term. *British Journal of Sociology* 44(3): 375-395.

Kumar, K. 1994. Civil Society again: a reply to Christopher Bryant's 'Social self-organization, civility and sociology'. *British Journal of Sociology* 45(1): 127-131.

Kuznetsov, E. 1990. The independent peace movements in the USSR, in V. Tismaneanu (ed.) *In Search of Civil Society: Independent Peace Movements in the Soviet Bloc.* London: Routledge.

Lane, D. 1996. *The Rise and Fall of State Socialism.* Cambridge: Polity.

Láng-Pickvance, K., Manning, N. and Pickvance, C. (eds) 1997. *Environmental and Housing Movements: Grassroots Experience in Hungary, Estonia and Russia.* Aldershot: Avebury.

Láng-Pickvance, K., Manning, N. and Pickvance, C. 1999. *Citizen Action in Eastern Europe.* Oxford University Press, Oxford, forthcoming.

Láng-Pickvance, K. *See also* Pickvance, K.

Lenin, V. I. 1969. 'Speech to the Moscow party workers' report on the attitude of the proletariat to the bourgeois democrats, in *Collected Works, Vol. 28.* Moscow: Foreign Languages Publishing House.

Lentin, A. 1973. *Russia in the Eighteenth Century.* London: Heinemann.

Levi, A. 1966. The Evolution of the Soviet System, in Z. Brzezinski (ed.) *Dilemmas of Change in Soviet Politics.* New York: Columbia University Press.

Lewin, M. 1968. *Russian Peasants and Soviet Power.* Evanston: University of Northern Illinois Press.

Lewin, M. 1985. *The Making of the Soviet System.* London: Methuen.

Lewin, M. 1988. *The Gorbachev Phenomenon: a Historical Interpretation.* Berkeley: University of California Press.

Liebman, M. 1970. *The Russian Revolution.* New York: Vintage Books.

Lincoln, W. B. 1989. *Red Victory: A History of the Russian Civil War.* London: Simon & Schuster.

Lincoln, W. B. 1990. *The Great Reforms, Autocracy, Bureaucracy, and the Politics of Change in Imperial Russia.* Evanston: University of Northern Illinois Press.

Lippay, Z. 1919. *The Hungarian Landed Middle Class and the Public Life.* Budapest: Franklin.

Littlejohn, G. 1984. *A Sociology of the Soviet Union.* London: Macmillan.

Lomax, B. 1976. *Hungary 1956.* London: Allison & Busby.

Mackenzie, D. 1977: *A History of Russia and the Soviet Union.* London: Dorsey Press.

Makkai, L. 1990. The foundation of the Hungarian Christian state, 950-1196. in F. P. Sugár (ed.) *A History of Hungary.* London: I. B. Tauris.

Mandel, E. 1989. *Beyond Perestroika: the Future of Gorbachev's USSR,* translated by G. Fagan. London: Verso.

Manicas, P.T. 1989. *War and Democracy.* Oxford: Basil Blackwell.

Manning, R. T. 1982. *The Crisis of the Old Order in Russia.* Princeton, N.J.: Princeton University Press.

Marx, K. 1973. *Grundrisse.* Harmondsworth: Penguin.

Marx, K. and Engels F. 1975. *Collected Works 3.* London: Lawrence and Wishart.

Marples, D. R. 1991. The greening of Ukraine: ecology and the emergence of Zelenyi Svit, 1986-1990, in J. Butterfield and J. B. Sedaitis (eds) *Perestroika from Below: Social Movements in the Soviet Union.* Boulder: Westview.

Massie, R. K. 1981. *Peter the Great.* London: Gollancz.

Mawdsley, E. 1987. *The Russian Civil War.* London: Allen & Unwin.

McAuley, M. 1992. *Soviet Politics, 1917-1991.* Oxford: Oxford University Press.

McNeal, R. H. 1975. *The Bolshevik Tradition.* Englewood Cliffs, N.J.: Prentice Hall.

Medvedev, R. A. 1980. *On Soviet Dissent.* (Interviews with P. Ostelino), edited by G. Saunders. New York: Columbia University Press.

Meissner, B. 1969. Totalitarian Rule and Social Change, in Z. Brzezinski, Z. *Dilemmas of Change in Soviet Politics.* New York: Columbia University Press.

Melucci, A. 1989. *Nomads of the Present. Social Movements and Individual Needs in Contemporary Society.* London: Century Hutchinson.

Milbrath, L. W. and Goel, M. L. 1965. *Political Participation.* Chicago: Rand McNally.

Mishan, E. J. 1967. *The Costs of Economic Growth.* Harmondsworth: Pelican.

Nagy, Z. L. 1975. Revolution in Hungary, 1918-1919, E. Pamlényi (ed.) *A History of Hungary.* London: Collets.

Nechemias, C. 1991. The Prospects for a Soviet Women's Movement: Opportunities and Obstacles, in J. Butterfield and J.B. Sedaitis (eds) *Perestroika from Below: Social Movements in the Soviet Union.* Boulder: Westview.

Nove, A. 1989. *Stalinism and After.* London: Allen and Unwin.

Oberschall, A. 1973. *Social Conflict and Social Movements.* Englewood Cliffs, N.J.: Prentice Hall.

Offe, C. 1984 *Contradictions of the Welfare State.* Cambridge, Mass.: MIT Press.

Offe, C. 1985. New social movements: changing boundaries of the political. *Social Research* 52: 817-868.

Offord, D. 1986. *The Russian Revolutionary Movement in the 1880s.* Cambridge: Cambridge University Press

Oliva, J. L. 1969. *Russia in the Era of Peter the Great.* Englewood Cliffs, N.J.: Prentice Hall.

Olson, M. 1965. *The Logic of Collective Action. Public Goods and the Theory of Groups.* Cambridge, Mass.: Harvard University Press.

Perrie, M. 1989. *Alexander II: Emancipation and Reform in Russia 1855-1881.* London: The Historical Association.

Pető, I. and Szakács, S. 1985. *A Hazai Gazdaság Négy Évtizedének Története 1945-1985: I. Az újjáépités és a tervutasitásos irányitás időszaka 1945-1968.* (The history of the four decades of the Hungarian economy 1945-85: I. The period of reconstruction and the command economy 1945-1968). Budapest: Közgazdasági és Jogi Könyvkiadó.

Pearson, T. 1989. *Russian Officialdom in Crisis: Autocracy and Self-Government, 1861-1900.* Cambridge: Cambridge University Press.

Perepjolkin, L. 1997. Environmental movements in Russia, in K. Láng-Pickvance, N. Manning and C. Pickvance (eds) *Environmental and Housing Movements: Grassroots Experience in Hungary, Estonia and Russia.* Aldershot: Avebury.

Pickvance, K. 1994. Towards a strategic approach to housing behaviour: a study of young people's strategies in South-East England (with C. Pickvance). *Sociology* 28(3): 657-677.

Pickvance, K. with Pickvance, C. 1995. The role of family help in the housing decisions of young people. *Sociological Review* 43(1):123-149.

Pickvance, K. 1996. Political participation and non-participation in Russia and Hungary. *Home Rule and Civil Society* 7(1): 101-123.

Pickvance, K. 1996. Popular protest and democracy: the Eastern European case, in C. Barker and M. Tyldesley (eds) *Alternative Futures and Popular Protest.* Manchester: The British Sociological Association Protest and Social Movements Group, Manchester Metropolitan University.

Pickvance, K. 1997. Environmental movements in Hungary and Russia: a comparative perspective. *European Sociological Review* 13(1): 1-25.

Pickvance, K. 1997. Environmental awareness in eastern Europe, in C. Barker and M. Tyldesley (eds) *Alternative Futures and Popular Protest.* Manchester: BSA Protest and Social Movements Group, MMU.

Pickvance, K. 1998. Democracy and opposition at grassroots level in eastern Europe: the case of environmental movements. *Sociological Review* 46(2): 187-207.

Pickvance, K. 1998. The diversity of post-socialist eastern European social movements, in P. Hamel and M. Mayer (eds) *Urban Movements and Global Processes.* London: Sage Publishers, forthcoming.

Pickvance, K. 1999. Popular Protest and Democracy in Eastern Europe. *International Journal of Comparative Sociology* 3(4), forthcoming.

Pickvance, K. *See also* Láng-Pickvance, K.

Piven, F.F. and Cloward, R.A. 1977. *Poor People's Movements: Why They Succeed, How They Fail.* New York: Pantheon.

Pokrovsky, M.N. 1970. Gentry capitalism, in T. Emmons (ed.) *Emancipation of the Russian Serfs.* London: Rinehart and Winston.

Pravda, A. 1979. Industrial workers: patterns of dissent, opposition and accommodation, in R. Tőkes (ed.) *Opposition in Eastern Europe*. London: Macmillan.

Pushkarev, S. 1963. *The Emergence of Modern Russia, 1801-1917*. New York: Holt, Rinehart and Winston.

Radice, H. 1981. The state enterprise in Hungary: economic reform and socialist entrepreneurship, in I. Jeffries (ed.) *The Industrial Enterprise in Eastern Europe*. Eastbourne: Praeger.

Raeff, M. 1973. The Enlightenment in Russia, in J. G. Garrard (ed.) 1973. *The Eighteenth Century in Russia*. Oxford: Clarendon Press.

Ránki, G. 1989. The role of Budapest in Hungary's economic development, in G. Ránki (ed.) *Hungarian History–World History*. Budapest: Akadémiai Kiadó.

Ránki, G. 1990. The Hungarian economy in the interwar years, in F. P. Sugár (ed.) *A History of Hungary*. London: I. B. Tauris.

Riasanovsky, N. V. 1984. *A History of Russia*. Oxford: Oxford University Press.

Richardson, D. 1995. The green challenge: philosophical, programmatic and electoral considerations, in D. Richardson D. and C. Rootes (eds) *The Green Challenge: The Development of Green Parties in Europe*. London: Routledge.

Rigby, T. H. 1990. *The Changing Soviet System*. Aldershot: Edward Elgar.

Robinson, G. T. 1970. The peasants in the last decades of serfdom, in T. Emmons (ed.) *Emancipation of the Russian Serfs*. London: Rinehart and Winston.

Rootes, C. 1982. Student Radicalism in France: 1968 and after, in Cerny P. (ed.) *Social Movements and Protest in France*. London: Frances Pinter.

Rootes, C. 1991. The New Politics and the New Social Movements in Britain, paper for the Political Studies Association Annual Conference.

Rootes, C. 1992. Political opportunity structures, political competition, and the development of social movements, paper presented at the First European Conference on Social Movements, Berlin.

Rootes, C. 1995. Environmental consciousness, institutional structures and political competition in the formation and development of Green parties, in R. Richardson and C. Rootes (eds) *The Green Challenge: The Development of Green Parties in Europe*. London: Routledge.

Rootes, C. 1997. Shaping collective action: structure, contingency and knowledge, in R. Edmondson (ed.) *The Political Context of Collective Action*. London: Routledge.

Rootes, C. and Davis H. (eds) 1994: *Social Change and Political Transformation*. London: UCL Press.

Rosenberg, W. G. 1987. Russian labour and Bolshevik power: social dimensions of protest in petrograd after October, in D. H. Kaiser (ed.) *The Workers' Revolution in Russia, 1917: The View from Below*. Cambridge: Cambridge University Press.

Rothschield, J. 1988. *East Central Europe Between the Two World Wars*. Seattle: Washington University Press.

Rozhdenstvenskij, S. V. 1963. Educational Reforms, in M. Raeff (ed.) *Peter the Great: Reformer or Revolutionary?* London: Heath.

Sablinsky, W. 1976. *The Road to Bloody Sunday*. Princeton, N.J.: Princeton University Press.

Sacharov, A. 1988. Speech. *New York Times*. 22 December.

Sakwa, R. 1990. *Gorbachev and his Reforms, 1985-1990*. London: Philip Alan.

Sakwa, R. 1993. *Russian Politics and Society.* London: Routledge.
Sakwa, R. 1994. The Russian elections of December 1993. *Europe-Asia Studies* 47: 195-228.
Schapiro, L. 1972. *Totalitarianism.* London: Pall Mall Press.
Schapiro, L. 1984. *1917 The Russian Revolutions.* Harmondsworth: Penguin Books.
Schmitter, P.C. 1982. *Patterns of Corporatist Policy Making.* London: Sage Publications.
Schöpflin, G. 1979. Opposition and para-opposition: critical currents in Hungary, 1968-78, in R. Tökés (ed.) *Opposition in Eastern Europe.* Baltimore: Johns Hopkins University Press.
Schöpflin, G. 1993. *Politics in Eastern Europe 1945-1992.* Oxford: Basil Blackwell.
Skocpol, T. 1979. *States and Social Revolutions: A Comparative Analysis of France, Russia and China.* Cambridge: Cambridge University Press.
Sedaitis, J. B. 1991. Worker activism: politics at the grass roots, in J. Butterfield and J. B. Sedaitis (eds) *Perestroika from Below: Social Movements in the Soviet Union.* Boulder: Westview.
Serge, V. 1972. *Year One of the Russian Revolution.* Harmondsworth: Penguin.
Seton-Watson, H. 1945. *Eastern Europe Between the Wars, 1918-1942.* Cambridge: Cambridge University Press.
Seton-Watson, H. 1967. *The Russian Empire 1801-1917.* Oxford: Oxford University Press.
Skilling, H. G. 1966. Communism and Czechoslovak Tradition. *Journal of International Affairs* 20(1): 118-137.
Skilling, H. G. 1972. Groups in Soviet politics: some hypotheses, in H. G. Skilling and F. Franklin (eds) *Interest Groups in Soviet Politics.* Princeton, N.J.: Princeton University Press.
Skilling, H. G. and Franklin, F. (eds) 1972 *Interest Groups in Soviet Politics.* Princeton, N.J.: Princeton University Press.
Slider, D. 1991. The first independent Soviet interest groups: unions and associations of cooperatives, in J. Butterfield and J. B. Sedaitis (eds) *Perestroika from Below: Social Movements in the Soviet Union.* Boulder: Westview.
Smelser, N. 1962. *Theory of Collective Behavior.* New York: The Free Press.
Smith, G. B. 1992. *Soviet Politics: Struggling with Change.* London: Macmillan.
Solomon, S. (ed.) 1983. *Pluralism in the Soviet Union: Essays in Honour of H.G. Skilling.* London: Macmillan.
Solomon, S. 1983. 'Pluralism' in Political Science: The Odyssey of a Concept, in S. Solomon (ed.) *Pluralism in the Soviet Union: Essays in Honour of H.G. Skilling.* London: Macmillan.
Soloveytchik, G. 1945. *Russia in Perspective.* London: Macdonald.
Soós, K. A. 1986. *Terv, kampány, pénz: Szabályozás és Konjuktúraciklusok Magyarországon és Jugoszláviában.* (Plan, campaign and money: regulations and cyclical changes in Hungary and Yugoslavia) Budapest: Közgazdasági és Jogi Könyvkiadó-Kossuth Kiadó.
Spector, I. 1965. *An Introduction to Russian History and Culture.* Princeton, N.J.: Princeton University Press.
Stewart, J. M. 1992. Air and water problems beyond the Urals, in J. M. Stewart (ed.)

The Soviet Environment: Problems, Policies and Politics. Cambridge: Cambridge University Press.

Sugár, F.P. 1990. The Principality of Transylvania, in F. P. Sugár (ed.) *A History of Hungary.* London: I. B. Tauris.

Suny, R. G. 1987. Revising the Old Story: the 1917 revolution in the light of new sources, in D. H. Kaiser (ed.) *The Workers' Revolution in Russia, 1917: The View from Below.* Cambridge: Cambridge University Press.

Szabó, M. 1993. *Alternativ Mozgalmak Magyarországon.* (Alternative movements in Hungary) Budapest: Gondolat Kiadó.

Szabó, M. 1994. Alternativ társadalmi mozgalmak Németországban és Magyarországon, (Alternative social movements in Germany and Hungary), *Politikatudományi Szemle,* 1: 96-103.

Szamizdat, '81-89: Válogatas a Hirmondó cimú folyóiratból. (Samizdat' 81-89, selected Articles from the journal Hirmondo) Budapest: AB-Beszelő Kft.

Szelényi, I. 1979. Socialist opposition in eastern Europe: dilemmas and prospects, in R. Tökes (ed.) *Opposition in Eastern Europe.* London: Macmillan.

Szelényi, I. and Konrád G. 1979. *The Intellectuals on the Road to Class Power.* Brighton: Harvester Press.

Szirmai, V. 1997. Protection of the environment and the position of green movements in Hungary, in K. Láng-Pickvance, N. Manning and C. Pickvance (eds) *Environmental and Housing Movements: Grassroots Experience in Hungary, Estonia and Russia.* Aldershot: Avebury.

Szücs, J. 1988. Three historical regions in Europe, in J. Keane (ed.) *Civil Society and the State.* London: Verso.

Swain, G. 1983. *Russian Social Democracy and the Legal Labour Movement.* London: Macmillan.

Tarrow, S. 1983. *Struggling to Reform. Social Movements and Policy Change during Cycles of Protest.* Cornell University, Western Societies Paper No.15.

Tarrow, S. 1989. *Democracy and Disorder: Protest and Politics in Italy, 1965-1975.* Oxford: Clarendon Press.

Tereshkova, V. 1987. Speech on 30 January 1987, reported by *Liberation* 3 February.

Thompson, 1990. *Russia and the Soviet Union.* Boulder: Westview.

Tilly, C. 1978. *From Mobilization to Revolution.* New York: Random House

Tismaneanu, V. 1988. *The Crisis of Marxist Ideology in Eastern Europe: The Poverty of Utopia.* London: Routledge.

Tismaneanu, V. (ed.) 1990. *In Search of Civil Society: Independent Peace Movements in the Soviet Bloc.* London: Routledge.

Tolz, V. 1990. *The USSR's Emerging Multi-Party System.* London: Praeger.

Tolz, V. 1994. Problems in building democratic institutions in Russia. *Radio Free Europe/Radio Liberty Research Report,* 3(9): 1-7.

Tökés, R. (ed) 1979. *Opposition in Eastern Europe.* Baltimore: Johns Hopkins University Press.

Turner, R. and Killian L. 1987. *Collective Behavior.* Englewood Cliffs, N.J., Prentice Hall.

Treadgold, D. W. (ed.) 1964. *The Development of the USSR.* Seattle: Washington University Press.

Troyat, H. 1961. *Daily Life in Russia Under the Last Tsar.* London: George Allen & Unwin.

Tsagalov, N. 1970. Systematic contradiction, in T. Emmons (ed.) *Emancipation of the Russian Serfs.* London: Rinehart and Winston.

Tucker, R. C. 1971. *The Soviet Political Mind.* New York: Norton.

Vajda, M. 1988. East-Central European perspectives, in J. Keane (ed.) 1988. *Civil Society and the State.* London: Verso.

Verba, S., Nie, N. H. and Kim, J. 1971. *The Modes of Democratic Participation.* London: Sage Publications.

Vernadsky, G. 1972. *A Source Book for Russian History from Early Times to 1917.* New Haven: Yale University Press.

Vida, L. 1995. Moszkvai hatalmi struktúra: Ostromállapot, (Moscow's power structure: a state of siege). *Heti Világgazdaság* 12(3)(817): 23-27.

Voronina, O. 1994. The Mythology of Women's Emancipation in the USSR as the Foundation for a Policy of Discrimination, in A. Posadskaya (ed.) *Women in Russia: A New Era in Russian Feminism.* London: Verso.

Waller, M. 1992. The dams on the Danube. *Environmental Politics,* 1: 121-143.

Waller, M. and Millard F. 1992. Environmental politics in eastern Europe. *Environmental Politics,* 1:159-185.

Walsh, W. B. 1963. *Readings in Russian History.* Volume I. Syracuse, N.Y.: Syracuse University Press.

White, S. 1991. *Gorbachev and After.* Cambridge: Cambridge University Press.

Williams, B. 1987. *The Russian Revolution, 1917-1921.* Oxford: Basil Blackwell.

Wilson, L. 1993. The Baikal Fund as an Environmental Pressure Group. Research Note. *Environmental Politics* 2(1): 63-80.

Weiner, D. R. 1988. *Models of Nature: Ecology, Conservation, and Cultural Revolution in Soviet Russia.* Bloomington: Indiana University Press.

Wood, A. 1979. *The Russian Revolution.* London: Longman.

Yaney, G. L. 1973. *The Systematization of Russian Government.* Urbana: University of Illinois Press.

Yanitsky, O. 1993a. *Russian Environmentalism: Leading Figures, Facts, Opinions.* Moscow: Mezhdunarodyje Otnoshenija Publishing House.

Yanitsky, O. 1993b. Environmental initiatives in Russia: east-west comparisons, in A. Vári A. and P. Tamás (eds) *Environment and Democratic Transition.* Dordrecht: Kluwer.

Yanitsky, O. and Khaly, I. A. 1993. *Invajronmental'noje Dvizhenie i Invajromental'naja Politika v Rossii.* (Environmental movement and environmental politics in Russia). Milieukunde: Universiteit van Amsterdam.

Zald, M. N. and Garner, R. A. 1987. Social movement organizations: growth, decay and change, in M. N. Zald and J. D. McCarthy (eds) *Social Movements in an Organizational Society: Collected Essays.* New Brunswick and Oxford: Transaction Books.

Ziegler, C. E. 1991. Environmental politics and policy under perestroika, in J. Butterfield and J. B. Sedaitis (eds) *Perestroika from Below: Social Movements in the Soviet Union.* Boulder: Westview.

Zimányi, V. 1987. *Economy and Society in 16th and 17th Century Hungary.* Budapest: Akadémiai Kiadó.

List of Interviews Quoted in the Study
(some of these are fictitious names in order to protect the real identity of the informers)

Bálint, Csaba
Bihari, Katalin
Boykov, Igor
Cherkasova, Elena
Ecopolis
Fejtö, Julia
Grakovich, Nikolay
Hársfalvi, Ágnes
Hollán, József
Kántor, Judit
Kékessy, Olga
Kemény, Kalman
Klimov, Piotr
Kovács, Judit
Kuranov, Mihail
Kreidlin, Anton
Makagonov, Ivan
Mizsei, József

Nagy, Andrea
Popov, Yuriy
Rubinchik, Ira
Rubenchik, Lubov'
Salgó, Lajos
Sárossy, Béla
Shalimov, Ovsey
Sokolov, Andrey
Strigulian, Sergey
Szalai, Irén
Szücs, Gábor
Tsyplenkov, Andrey
Utassy, Éva
Varjú, Margit
Vorobiev, Ivan
Zabelin, Igor
Zaykonova, Natasha

Index